THE ROLE OF THEORY
IN ADVANCING 21ST-CENTURY BIOLOGY

Catalyzing Transformative Research

Report of the Committee on Defining and Advancing the Conceptual Basis
of Biological Sciences in the 21st Century

Board on Life Sciences

Division on Earth and Life Studies

NATIONAL RESEARCH COUNCIL
OF THE NATIONAL ACADEMIES

THE NATIONAL ACADEMIES PRESS
Washington, D.C.
www.nap.edu

THE NATIONAL ACADEMIES PRESS   500 Fifth Street, NW   Washington, DC 20001

NOTICE: The project that is the subject of this report was approved by the Governing Board of the National Research Council, whose members are drawn from the councils of the National Academy of Sciences, the National Academy of Engineering, and the Institute of Medicine. The members of the committee responsible for the report were chosen for their special competences and with regard for appropriate balance.

This material is based on work supported by the National Science Foundation under Grant No. DBI–0633909. Any opinions, findings, and conclusions or recommendations expressed in this material are those of the author(s) and do not necessarily reflect the views of the National Science Foundation, nor does mention of trade names, commercial products, or organizations imply endorsement by the U.S. government.

International Standard Book Number-13:   978-0-309-11249-9 (Book)
International Standard Book Number-10:   0-309-11249-4 (Book)
International Standard Book Number-13:   978-0-309-11250-5 (PDF)
International Standard Book Number-10:   0-309-11250-8 (PDF)
Library of Congress Control Number: 2007940783

Additional copies of this report are available from the National Academies Press, 500 Fifth Street, NW, Lockbox 285, Washington, DC 20055; (800) 624-6242 or (202) 334-3313 (in the Washington metropolitan area); Internet, http://www.nap. edu.

Cover: Design by Francesca Moghari; artwork by Nicolle Rager Fuller (*www.sayo-art.com*).

THE NATIONAL ACADEMIES
Advisers to the Nation on Science, Engineering, and Medicine

The **National Academy of Sciences** is a private, nonprofit, self-perpetuating society of distinguished scholars engaged in scientific and engineering research, dedicated to the furtherance of science and technology and to their use for the general welfare. Upon the authority of the charter granted to it by the Congress in 1863, the Academy has a mandate that requires it to advise the federal government on scientific and technical matters. Dr. Ralph J. Cicerone is president of the National Academy of Sciences.

The **National Academy of Engineering** was established in 1964, under the charter of the National Academy of Sciences, as a parallel organization of outstanding engineers. It is autonomous in its administration and in the selection of its members, sharing with the National Academy of Sciences the responsibility for advising the federal government. The National Academy of Engineering also sponsors engineering programs aimed at meeting national needs, encourages education and research, and recognizes the superior achievements of engineers. Dr. Charles M. Vest is president of the National Academy of Engineering.

The **Institute of Medicine** was established in 1970 by the National Academy of Sciences to secure the services of eminent members of appropriate professions in the examination of policy matters pertaining to the health of the public. The Institute acts under the responsibility given to the National Academy of Sciences by its congressional charter to be an adviser to the federal government and, upon its own initiative, to identify issues of medical care, research, and education. Dr. Harvey V. Fineberg is president of the Institute of Medicine.

The **National Research Council** was organized by the National Academy of Sciences in 1916 to associate the broad community of science and technology with the Academy's purposes of furthering knowledge and advising the federal government. Functioning in accordance with general policies determined by the Academy, the Council has become the principal operating agency of both the National Academy of Sciences and the National Academy of Engineering in providing services to the government, the public, and the scientific and engineering communities. The Council is administered jointly by both Academies and the Institute of Medicine. Dr. Ralph J. Cicerone and Dr. Charles M. Vest are chair and vice chair, respectively, of the National Research Council.

www.national-academies.org

MARILEE K. SHELTON-DAVENPORT, Senior Program Officer
EVONNE P.Y. TANG, Senior Program Officer
REBECCA WALTER, Program Assistant
ROBERT T. YUAN, Senior Program Officer

Acknowledgments

This report is a product of the cooperation and contributions of many people. The members of the committee thank all of the speakers who briefed the committee. (Appendix C presents a list of presentations to the committee.)

This report has been reviewed in draft form by persons chosen for their diverse perspectives and technical expertise in accordance with procedures approved by the National Research Council's Report Review Committee. The purpose of this independent review is to provide candid and critical comments that will assist the institution in making its published report as sound as possible and to ensure that the report meets institutional standards of objectivity, evidence, and responsiveness to the study charge. The review comments and draft manuscript remain confidential to protect the integrity of the deliberative process. We wish to thank the following people for their review of this report:

Douglas Cook, University of California, Davis
Daniel Dennett, Tufts University, Medford, Massachusetts
Robert Full, University of California, Berkeley
Lou Gross, University of Tennessee, Knoxville
Alan Hastings, University of California, Davis
J. Woodland Hastings, Harvard University, Cambridge, Massachusetts
Douglas Lauffenburger, Massachusetts Institute of Technology
Simon Levin, Princeton University
Kenneth Nealson, University of Southern California, Los Angeles

Jeffrey Platt, Mayo Clinic, Rochester, Minnesota
Tom Pollard, Yale University, New Haven, Connecticut
Rino Rappuoli, Chiron Corp, Siena, Italy
Hudson Kern Reeve, Cornell University, Ithaca, New York
Gene Robinson, University of Illinois, Urbana-Champaign
Michael Ryan, University of Texas, Austin
Kevin Strange, Vanderbilt University
Keith Yamamoto, University of California, San Francisco

Although the reviewers listed above provided constructive comments and suggestions, they were not asked to endorse the conclusions or recommendations, nor did they see the final draft of the report before its release. The review of this report was overseen by Dr. May Berenbaum, University of Illinois. Appointed by the National Research Council, Dr. Berenbaum was responsible for making certain that an independent examination of this report was carried out in accordance with institutional procedures and that all review comments were carefully considered. Responsibility for the final content of this report rests entirely with the author committee and the institution.

Contents

Summary

From microorganisms to whales, from single cells to complex organisms, from plants to animals to fungi, from body plans to behavior, the diversity of life is amazing. Living organisms have a profound impact on our physical world of ocean, landscape, and climate; around us is a multitude of diverse ecosystems that provide a livable environment and many valuable resources. The study of life—biology—is a multifaceted endeavor that uses observation, exploration, and experiments to gather information and test hypotheses about topics ranging from climate change to stem cells. The field of biology is so diverse that it can sometimes be hard for one individual to keep its breadth in mind while contemplating a particular question.

This study was initiated at the request of, and with the sponsorship of, the National Science Foundation. It was conceived as a new approach to a question that has been asked before: What is the future of biology? In 1989 the National Research Council released a report on this topic entitled *Opportunities in Biology*. Over 400 pages long and four years in the making, the report provided a detailed snapshot of the state of biology at that time. Eleven different panels detailed the opportunities awaiting the rapidly diversifying field of biology. Reading the report today, the excitement of that time is palpable. Section after section describes new technologies and promises new discoveries. Each section focuses on a different subdiscipline of biology.

This report takes a different approach by looking for commonalities across subdisciplines. The committee was charged with examining the role of concepts and theories in biology, including how that role might differ across various subdisciplines. One facet of that examination was to consider the role of the concepts and theories in driving scientific advances and to make recommendations about the best way to encourage creative, dynamic, and innovative research in biology. The charge was to focus on basic biology, not on biomedical applications.

At the first committee meeting, to begin identifying the theoretical foundations of biology, each committee member discussed the theories and concepts underlying his or her particular area of research and addressed how those theories and concepts might connect across the field of biology. The talks demonstrated that biologists from all subdisciplines base their work on rich theoretical foundations, albeit of very diverse kinds. They highlighted the varied extent to which theories are an explicit focus of attention and discussion. For example, cell theory underpins much research, but the theory itself is rarely the topic of explicit attention in the research literature.

The committee concluded that a more explicit focus on theory and a concerted attempt to look for cross-cutting issues would likely help stimulate future advances in biology. To illustrate this point, the committee chose seven questions to examine in detail. The list of questions is not comprehensive but rather illustrative. The questions, as shown below, were chosen to show that a focus on theory could play a role in helping to address many different types of interesting and important questions at many different levels.

1.   Are there still new life forms to be discovered?

New organisms continue to be discovered, some in environments that were once thought incompatible with life. How many new life forms remain to be discovered? What additional strategies for movement, sensation, and chemical synthesis will be found? How diverse are the variations on the patterns of development of organisms' body plans? How do complicated communities of different organisms affect each other's evolution and what can be learned from the diversity of social organizations that have evolved in different species? How is diversity encouraged and limited by environment? For billions of years, life was exclusively microbial—to what degree can a better understanding of that early evolution change our understanding of the present microbial world, which is turning out to be vastly more diverse than ever imagined, and the processes that underlie all life forms?

The diversity of life presents a huge challenge to biologists but also a virtually limitless opportunity. Both the unity and the diversity of life are explained by the theory of evolution: All life forms share many characteristics because all are descended from a common ancestor and life has become diverse through billions of years of descent with modification. However, the extent and implications of all that diversity are not yet fully understood. An enormous amount of productive research has demonstrated many mechanisms by which evolution leads to diversity. However, much remains to be described and explained. There is need for further theoretical insight into how diversity is generated and maintained, not to mention understanding the implications of losses of diversity. These are exciting challenges.

2.   What role does life play in the metabolism of planet Earth?

Diverse as life is, the metabolic pathways that support it are, perhaps surprisingly, quite well conserved and are based on just a few basic strategies. These metabolic pathways, which are the means by which organisms acquire the energy and material components they need to survive and reproduce, have a profound global impact as living organisms form part of global geochemical cycles. The Earth today has been shaped in many ways by metabolic processes, which are key molecular processes at the cellular level as well. Understanding the evolution of these pathways, how they integrate, and how living systems are coupled to environmental conditions is a profoundly important question to several areas of biology and on many scales of time and space.

3.   How do cells really work?

The living cell is a marvel, containing thousands of interlocked chemical reactions that harvest energy from the environment, synthesize thousands of different chemicals, manage waste, and recycle components. Ultimately, the cell makes a copy of itself. No human factory can rival the cell's compact and coordinated productivity. Only a fraction of its pathways can be reproduced in the test tube. The laws of physics and chemistry apply, of course, to all living organisms. However, most life processes are maintained far from chemical and thermodynamic equilibrium. Thus, understanding how chemical reactions take place in the crowded and highly organized molecular environment of the cell, or how physical variables like temperature and concentration gradients affect and are affected by living processes (for example, during development, or in the cell cycle or circadian cycle, when the instructions encoded by DNA are manifested in physical processes), is a major challenge of biological research. The interfaces between some current research areas of physics, chemistry, and biology that elucidate these questions are expected to be very fruitful.

4.   What are the engineering principles of life?

DNA is made up of nucleotides, proteins of amino acids. Organisms contain many types of cells, ecosystems many different species. The hierarchical organization of building blocks at different scales is a common theme in biology, whose evolution is not fully understood. Complicated systems at every scale are made up of simpler modules that vary in definable ways and combine in ways that result in structures capable of much more than the individual parts. This characteristic of complicated structures, functions, and behaviors arising from the combination of simple parts represents an almost universal theme in biology. Furthermore, across all scales of biology, from subcellular circuits to ecosystems, many biological systems demonstrate "robustness": in other words, they continue to function despite defective parts or changes in the environment. Like the workings of the living cell, this robustness is a biological phenomenon that has evolved through variation and selection and that human engineers would be proud to duplicate. Understanding the principles by which modules combine to create systems with particular properties (another useful, cross-cutting concept) will undoubtedly result in theoretical insights that would apply across biological scales from the molecular to the ecosystem—and perhaps provide valuable lessons for human efforts in design and engineering.

5. What is the information that defines and sustains life?

The power of the computer rests in its ability to represent an immense range of phenomena in digital form that can then be manipulated. Many of the characteristics of life can similarly be represented as flows of information, as it is striking that all living organisms and communities of organisms are able to sense, process, remember, and respond to many different kinds of external and internal stimuli that can be conceptualized as information. Evolution, for example, can be viewed as a process whereby selection of variant genomes is affected by the information provided by the environment. In this view, the information defining the relevant environmental variables is partly encoded in the genome of the adapted organisms by the process of selection, and evolution is thus a process of selective memory in molecular form stored in the genomes of living organisms. The complexity of biological systems can be described using the ideas of information science, but there are deeper conceptual problems in making full use of those concepts of information that were developed for engineering and physics, where they are used in pattern recognition, communications, and thermodynamics. In biological systems, information is intimately dependent on context, making it difficult to apply the concept of information in ways that do not oversimplify complicated biological phenomena. Thus, further development of both concepts and tools will be required to realize the potential of this powerful conceptual point of view.

6. What determines how organisms behave in their worlds?

Organisms as diverse as bacteria and humans possess the ability to respond to their environments and to shape their behaviors in response to specific environmental variables. Understanding how organisms live requires determining the rules that govern how organisms behave in their world, how they sense their environments, and how they use this information to change their behavior. It is important to remember that organisms do not simply wait passively for information from their environments. Their physiology is internally generated, by genetically determined rules, and input from the environment is used to alter the behavior of the organism. In addition, much behavior is generated to actively explore the environment in search of specific sensory signals. For example, bacteria have receptor proteins that allow them to sense concentrations of chemicals in their environment and use these gradients to govern their movements. The integration of sensory information into a form that can be processed by the organism, the nature of the processing machinery, the influence of the internal states of the organism, the influence of the experience on the future states of the organism, memory mechanisms, and many other issues have direct relevance to many different biological regimes, scales, and kinds of organisms. There is a remarkable potential for finding commonalities amid the diversity addressed by this question.

7. How much can we tell about the past—and predict about the future—by studying life on Earth today?

The ability of living systems to pass on the directions for reproducing themselves and for surviving in the environments where those offspring will find themselves is fundamental to the living state, and it is more than a loose metaphor to say that organisms' genomes represent an imprint of past environmental conditions, history, and the selection pressures on the ancestors of organisms. The sequences of the genome are not the only records of past conditions; the ways in which those sequences are put to use are also affected by other past conditions that are carried forward by living systems—from stable physiological states, to imprinted DNA that modifies gene expression, to memories stored in the brain and nervous system, and behaviors remembered and taught to descendants. New mechanisms and new applications of this common ability of living things to record information about the past in some physical, molecular form continue to emerge. Thus, the commonality and diversity in the ways in which organisms represent and use this kind of information are very promising and very challenging frontiers for future research. The record of the past is imprinted in both the fossil record and the DNA of today's living world. Whatever life on Earth looks like 1 million years from now, it will evolve from what is currently alive. If scientists truly understood how current organisms and environments interact to produce future generations, could the course of evolution be predicted?

FINDINGS AND RECOMMENDATIONS

Of course, it is impossible to cover all of biology in so short a report. If the average freshman biology textbook needs hundreds of pages to cover the basics, a mere seven questions cannot possibly introduce even a fraction of the exciting and innovative biology research that is currently underway. The questions are meant to be illustrative, not all-inclusive, and should be read not as a guide to the most important or promising areas for future emphasis but as several examples of the way that concepts and theories can connect the different areas of biology. After exploring this set of seven questions, the committee came to consensus on several findings and recommendations that flow from the idea of looking at cross-cutting issues in biology with an eye to the role of theory.

Finding 1

Biological science can contribute to solving societal problems and to economic competitiveness. Basic and applied research targeted toward a particular mission is one way to accomplish this important goal. However, increased investment in the development of biology's fundamental theoretical and conceptual basis is another way to reap practical benefits from basic biological research. Theory is an integral part of all biological research, but its role is rarely explicitly recognized.

The living world presents a vast reservoir of biological solutions to many practical challenges, and biological systems can inspire innovation in many fields. The many ways that basic biological research contributes to medicine are very familiar, but basic biology can also contribute to advances in fields as diverse as food, fishery, and forest production; pest management; resource management; conservation; transportation; information processing; materials science; and engineering. Biological research breakthroughs, therefore, have the potential to contribute to the solution of many pressing problems, including global warming, pollution, loss of biodiversity, fossil fuel dependence, and emerging infectious diseases.

As the many examples in this report attest, biology is characterized by unity and diversity. There is unity because many biological processes have been preserved through evolution. There is also diversity because natural selection has led to many innovative solutions to the practical problems that living organisms have encountered over billions of years. Therefore, discoveries about a particular organism, sensory pathway, or regulatory network can have immediate applications throughout biology, and the transformative insight that provides the most direct path to a practical solution may arise in a seemingly unrelated research area. Giving explicit recognition to the role of theory in the practice of biology and increasing support for the theoretical component of biology research are ways to help make such connections and thus leverage the value of basic biological research.

The extent of life's diversity has not yet been plumbed, and many biological processes are understood only imperfectly. New tools and computational capabilities are improving biologists' ability to study complex phenomena. Tying together the results of research in the many diverse areas of biology requires a robust theoretical and conceptual framework, upon which a broad and diverse research portfolio of basic biological investigations can be based. The impact of biology on society could be enhanced if discovery and experimentation are complemented by efforts to continuously enrich biology's fundamental theoretical and conceptual basis.

Recommendation 1

Theory, as an important but underappreciated component of biology, should be given a measure of attention commensurate with that given other components of biological research (such as observation and experiment). Theoretical approaches to biological problems should be explicitly recognized as an important and integral component of funding agencies' research portfolios. Increased attention to the theoretical and conceptual components of basic biology research has the potential to leverage the results of basic biology research and should be considered as a balance to programs that focus on mission-oriented research.

Finding 2

Biologists in all subdisciplines use theory but rarely recognize the integral and multifaceted role that theory plays in their research and therefore devote little explicit attention to examining their theoretical and conceptual assumptions. Major advances in biological knowledge come about through the interplay of theoretical insights, observations, and key experimental results and by improvements in technology that make new observations, experiments, and insights possible. The fragmentation of biology into many subdisciplines means both that the mix of these components can differ dramatically from one area to another and that the development of theoretical insights that cut across subdisciplines can be difficult. It is the committee's opinion that all subdisciplines of biology would benefit from an explicit examination of the theoretical and conceptual framework that characterizes their discipline.

Recommendation 2

Biology research funding portfolios should embrace an integrated variety of approaches, including theory along with experiment, observation, and tool development. Biologists in all subdisciplines should be encouraged to examine the theoretical and conceptual framework that underlies their work and identify areas where theoretical advances would most likely lead to breakthroughs in our understanding of life. Workshops sponsored by funding agencies or scientific societies would be one way to facilitate such discussions. The theoretical and conceptual needs identified by such subdisciplinary workshops should then be integrated into the funding programs for those subdisciplines. It would also be worthwhile to sponsor interdisciplinary workshops to identify theoretical and conceptual approaches that would benefit several subdisciplines.

Finding 3

New ways of looking at the natural world often face difficulty in acceptance. Challenges to long-held theories and concepts are likely to be held to a higher standard of evidence than more conventional proposals. Proposals that break new ground can face difficulty in attracting funding, for example those that cross traditional subdisciplinary boundaries, take a purely theoretical approach, or have the potential to destabilize a field by challenging conventional wisdom. Such proposals are likely to be perceived as "high-risk" in that they are likely to fail. However, their potential for high impact warrants special attention. Successfully determining which of them deserve funding will require input from an unusually diverse group of reviewers.

Recommendation 3

Some portion of the basic research budget should be devoted to supporting proposals that are high risk and do not fall obviously into present funding frameworks. One possibility is to initiate a program specifically for such "high-risk/high-impact" proposals—whether they are purely theoretical, cross-disciplinary, or unconventional. Another is to encourage program officers to include some proportion of such proposals in their portfolios. A third is to provide unrestricted support to individuals or teams of scientists who have been identified as particularly innovative. Evaluation of these proposals should be carefully designed to ensure that reviewers with the requisite technical, disciplinary, and theoretical expertise are involved and that they are aware of the goal of supporting potentially consensus-changing research. Proposals that challenge conventional theory require not only that the originality and soundness of the theoretical approach be evaluated but also that the biological data being used are appropriate and the question being asked is significant.

Finding 4

Technological advances in arrays, high-throughput sequencing, remote sensing, miniaturization, wireless communication, high-resolution imaging, and other areas, combined with increasingly powerful computing resources and data analysis techniques, are dramatically expanding biologists' observational, experimental, and quantitative capabilities. Questions can be asked, and answered, that were well beyond our grasp only a few years ago. It is the committee's contention that an increased focus on the theoretical and conceptual basis of biology will lead to the identification of even more complex and interesting questions and will help biologists conceive of crucial experiments that cannot yet be conducted. Biologists' theoretical framework profoundly affects which tools and techniques they use in their work. All too frequently, experimental and observational horizons are unconsciously limited by the technology that is currently available. Advances in technology and computing can provide biologists with many new opportunities for experimentation and observation.

For many of the multiscale questions raised in this report, there is a strong need for teams of biologists, engineers, physicists, statisticians, and others to work together to solve cross-disciplinary problems. The interaction and collaboration of biologists with physicists, engineers, computer scientists, mathematicians, and software designers can lead to a dynamic cycle of developing new tools specifically to answer new questions rather than limiting questions to those that can be addressed with current technology. The growing role and shortening life cycle of technology mean that biologists will have to become ever more adept in the use of new equipment and analysis techniques. Understanding the capabilities, and especially the limitations, of new instruments so that experiments are designed properly and results interpreted appropriately will be important in more and more areas of biology.

Because the potential benefits of more precise and rapid measurements of biological phenomena are so high, it will be important for biologists to be aware of both instrumentation capabilities in the physical and engineering sciences and theoretical advances in physics, chemistry, and mathematics that could be integrated into biological research. Conversely, if researchers outside biology are aware of the kinds of questions biologists are now asking, they can use their techniques, instruments, and approaches to advance biological research. Close collaboration between biologists and researchers in other fields has great promise for leveraging the value of discoveries and theoretical insights arising from basic biological research.

Recommendation 4

In order to gain the greatest possible benefit both from discoveries in the biological sciences and from new technological capabilities, biologists should look for opportunities to work with engineers, physical scientists, and others. Funding agencies should consider sponsoring interdisciplinary workshops focused on major questions or challenges (such as understanding the consequences of climate change, addressing needs for clean water, sustainable agriculture, or pollution remediation) to allow biologists, scientists from other disciplines, and engineers to learn from each other and identify collaborative opportunities. Such workshops should be designed to consider not just what is possible with current technology but also what experiments or observations could be done if technology were not an obstacle. Opportunities for biologists to learn about new instrumentation and to interact with technology developers to create new tools should be strongly supported. One possible approach would be the creation of an integrative institute focused on bioinstrumentation, where biologists could work in interdisciplinary teams to conceive of and develop new instrumentation. The National Center for Ecological Analysis and Synthesis and the National Evolutionary Synthesis Center could serve as models for the development of such an institute.

Finding 5

To get the most out of large and diverse data sets, these will need to be accessible and biologists will have to learn how to use them. While technology is making it increasingly cost-effective to collect huge volumes of data, the process of extracting meaningful conclusions from those data remains difficult, time-consuming, and expensive. Theoretical approaches show great promise for identifying patterns and testing hypotheses in large data sets. It is increasingly likely that data collected for one purpose will have relevance for other researchers. Therefore, the value of the data collected will be multiplied if the data are accessible, organized, and annotated in a standardized way. While it is somewhat new to many areas of biology, other fields—like astronomy and seismology—that create massive data sets rely on theory to guide pattern detection and to direct *in silico* experimentation and modeling. Getting the most out of the extensive biological data that can now be collected will increasingly require that biologists broadly develop those skills and collaborate with mathematicians, computer scientists, statisticians, and others. This process of building community databases is well underway in many areas of biology, genomics being a prominent example, but the specialized databases developed by one research community may be unknown or inaccessible to researchers in other fields. Significant resources are needed to maintain, curate, and interconnect biological databases.

Recommendation 5

Attention should be devoted to ensuring that biological data sets are stored and curated to be accessible to the widest possible population of researchers. In many cases, this will require standardization. Providing opportunities for biologists to learn from other disciplines that routinely carry out theoretical research on diverse data sets should also be explicitly encouraged.

1

Introduction

THE TANGLED WEB OF BIOLOGICAL SCIENCE

The diverse living things of our world are endlessly fascinating. Living organisms have a profound impact on the physical world of ocean, landscape, and climate, and around us is a multitude of diverse ecosystems that provide us with a livable environment and many valuable resources. There is a vast array of interactions among living things, including those that characterize human society and the relationship between humans and the rest of the living world. The practice of biological science takes many forms, with observation, exploration, and experiment combining in many ways to gather information and test hypotheses. The means by which these practices are actually carried out is profoundly affected by the technologies available, with new tools regularly opening up new realms to experimentation, observation, analysis, and novel conceptual insight. Both biologists and nonbiologists occasionally caricature biology in these terms—a science dedicated to endless observation, collection, and testing, leading to a snowballing accumulation of facts. Life is so complex and science has examined such a small fraction of its diversity that it seems reasonable to think that a great deal more data are needed before unifying theories can emerge that explain life in all its diversity. One goal of this report is to illustrate that we need not, and do not, sit and wait for theory to emerge as the end game of biological research. Theory is already an inextricable thread running throughout the practice of biology, as it is in all science. Biologists choose where to observe, what tool to use, which experiment to do, and how to interpret their results on the basis of a rich theoretical and conceptual framework. Biologists strive to discern patterns, processes, and relationships in order to make sense of the seemingly endless diversity of form and function. Explanatory theories are critical to making sense of what is observed—to order biological phenomena, to explain what is seen and to make predictions, and to guide observation and suggest experimental strategies. Because the living world is so complex, biological theory is also exceptionally rich and varied.

Science is facts; just as houses are made of stones, so is science made of facts; but a pile of stones is not a house and a collection of facts is not necessarily science

—Henri Poincare, French mathematician and physicist
(1854-1912)
(Mackay, 1991)

What makes the house of biology from the pile of stone facts is the theoretical component.

THE ORIGIN OF THIS REPORT

In 1989 the National Research Council released a report entitled *Opportunities in Biology*. Over 400 pages long and four years in the making, the report provides a detailed snapshot of the state of biology at that time. Eleven different panels described the opportunities awaiting the rapidly diversifying field of biology. Reading the report today, the excitement of that time is palpable. Section after section describes new technologies and promises new discoveries. The technologies span many levels, from the molecular—DNA sequencing technology had recently progressed from manual to automatic—to the ecological, as robotic arms and free-ranging robots were dramatically expanding the ability of deep-sea submersibles to survey and sample the ocean floor. Nearly 20 years later, it appears that in many respects the authors of that report underestimated the power of the new technologies they described. In 1989 a total of 15 million nucleotides of DNA sequence had been determined. The latest generation of sequencing machines can sequence more than 100 million nucleotides per day. Satellites allow biologists to examine changes in landscapes on an ever finer scale and to track wildlife remotely, while the World Wide Web allows them to retrieve and share their data instantly.

The productivity of biological research since 1989 has been extraordinary. At the same time, the explosion of new biological information has consequences. Individual scientists can now collect data on a scale and at a level of detail that surpass any individual's capacity to sift through, analyze, and interpret all that can be collected. Ever more sophisticated experimental approaches to deciphering how the endless variety of biological systems function opens up a universe of potential experiments so vast that no number of biologists even with unlimited resources could undertake them all. In fact, so much information is accumulating, on so many different biological systems, that it has become impossible for any one biologist to stay abreast of all the advances being made even within one subfield, much less throughout all of biology. There is a growing sense that the ability to collect such a large amount of data, while welcome, also poses new challenges: Are the data already collected adequately organized and accessible, and how can the constant influx of new data be put to best use? How do we decide what experiments to do, which data to collect? There is tremendous potential that new technologies and computational approaches will allow biologists to ask and answer questions that were unmanageable in the past and that chemically and physically reasonable explanations for many complicated biological phenomena will continue to emerge. It is worth considering whether we have the tools and resources necessary to identify potentially unifying themes or organizing principles. A sequel to the 1989 report examining in that same spirit today's "Opportunities in Biology" could easily require 800 pages and 22 subcommittees and would identify hundreds of exciting potential areas for biological discovery. Continuing on the ever-widening research path illuminated in the 1989 report would no doubt lead to great achievements—the record of biological research over the last 20 years has been impressive. At the same time, this is an opportune moment to take stock of the field of biology and examine whether a different perspective is in order, one that might allow biological science to advance faster and contribute even more effectively to addressing the pressing needs of society.

Study Process

This project was initiated at the request of, and with the sponsorship of, the National Science Foundation. The committee first met to discuss its charge and goals in October 2006 and then held a workshop to gather additional input in December 2006. Subsequent meetings in the spring of 2007 were held to work on report writing.

The committee was charged to identify and examine the concepts and theories that form the foundation for scientific advancement in various areas of biology, including (but not limited to) genes, cells, ecology, and evolution. It was asked to assess which areas are "theory-rich" and which areas need stronger conceptual foundations for substantial advancement and to make recommendations as to the best way to encourage creative, dynamic, and innovative research in biology. Building on these results, the study was to identify major questions to be addressed by 21st-century biology. The project was to focus on basic biology, but not on biomedical applications. Questions that could be considered by the committee included:

- What does it mean to think of biology as a theoretical science?
- Is there a basic set of theories and concepts that are understood by biologists in all subdisciplines?
- How do biological theories form the foundation for scientific advancement?
- Which areas of biology are "theory-rich" and which areas need stronger conceptual foundations for substantial advancement?
- What are the best ways to bring about advances in biology?
- What are the grand challenges in 21st-century biology?
- How can educators ensure that students understand the foundations of biology?

At its first committee meeting, in order to identify common theories and concepts in biology, each committee member was asked to present the theories and concepts underlying his or her particular area of research and address how those theories and concepts might apply across biology in general. If the hope was that the talks would unearth a set of theories in each area of biology, sets that could then be compared to find commonalities and show which areas were particularly "theory-rich" and where theory was notably lacking, the result was quite different. The talks demonstrated that biologists from all subdisciplines base their work on rich theoretical foundations, albeit of very diverse types and mixtures. What became evident was the universality of the complex interaction between current theories, new observations and experimental evidence, and evolving technological capabilities. Those areas in which prevailing theory is being challenged through observation, experiment, and analysis are likely to be where the most interesting biology research is being done. This should not have been a surprise for this is a common phenomenon—the recognition that facts are accumulating that contradict the prevailing theoretical framework often characterizes highly active and exciting research and a field in which important changes are imminent. At its second meeting, the committee invited a diverse group of biologists, focusing especially on researchers in subdisciplines that were not represented on the committee, to give talks discussing the theories and concepts underlying their research. Again, the talks did not identify discrete sets of theories that characterized particular areas of research, with some areas having richer theory sets than others. The two sets of talks convinced the committee that identifying a list of concepts and theories that underlie different areas of biology, as requested in the first line of the Statement of Task, would not accurately represent the role of theory in biology. This is not to say that biology has no foundational theories that are accepted by all biologists; evolution, cell theory, and the role of DNA in inheritance certainly serve that unifying purpose. However, the committee did not find that each subdiscipline of biology has its own more or less well-developed set of foundational theories. The committee's assessment of which areas of biology were "theory-rich" and which areas needed stronger conceptual foundations for substantial advancement concluded that all areas of biology rest on a rich theoretical framework but that the range and types of theories in use were exceptionally diverse.

Despite the difficulty that the committee found in responding literally to the Statement of Task, the committee welcomed the opportunity to explore the integral role that theory plays in biology and to point out the ways that theory contributes to creative, dynamic, and innovative research in biology. The committee then decided to use a set of broad questions with relevance across many subdisciplines of biology to illustrate the role that theory now plays and might play even more prominently in the future. The goal was to choose questions that would illustrate the many connections across biological scales and subdisciplines, not to cover the field comprehensively nor to identify which new areas of research are the most important or promising. Inevitably, this approach precluded covering any area in depth and made it impossible to mention all of the many interesting and innovative areas of current biological research.

Where Do Transformative Insights Come From?

In the history of biology, one can identify many moments when our understanding of the living world was transformed. Some of these transformative moments have resulted from a deep insight that led to a major change in our theoretical framework. Other transformative moments were triggered by a key observation or experimental result, or by the invention of a new tool for making observations or doing experiments. None of these moments came about, though, without complex interaction among the many components that make up the practice of biology. Certainly one of the most transformative moments in biology was Darwin's exposition of the theory of evolution by natural selection. His insights have since inspired and elucidated more than a century of rewarding observation and experimentation, richly demonstrating how the process of evolution has resulted in so many diverse life forms, functions, and patterns. But what made possible the transformative moment that was Darwin's theoretical insight? First, an accumulation of facts (in the form of diverse fossil remains) emerged that were difficult to reconcile with the prevailing theory of a fixed and unchanging collection of species. Second, the collection and organization of hundreds of thousands of samples of biological specimens in the museums of Europe during the 19th century (made possible by improvements in navigational tools and motivated by a desire to catalog the diversity of creation), as well as Darwin's own observations and collection during his famous voyage—in other words, the curiosity-driven collection of data about the living world—provided the raw material that enabled Darwin's theoretical insight.

Another profoundly transformative moment, the elucidation of the structure of DNA, could not have happened before the key technological capability of X-ray diffraction was available. Together with the evidence that DNA was the critical substance that passed from generation to generation and that its four simple components were always found in a consistent ratio, Watson and Crick brought to their efforts a theoretical construct. (They had models of the diffraction patterns that helical molecules should produce.) The physical evidence provided by the X-ray scattering patterns that DNA was a molecular double helix was the final link that tied the theory and all the observations together, suggested molecular mechanisms of replication and inheritance, and gave rise to a transforming era in biology.

It is important to note that the tangle of facts, observation, experiment, theory, and technology has no particular beginning and certainly no end. At different times, one of these components may receive more emphasis, but major advances in modern biology have never been completed without all of them.

Despite the integral role that theory plays throughout the practice of biology, biologists rarely think of themselves as theoretical scientists. Part of the reason is that the word "theory" can be used to mean many different things, ranging from a mere hunch to a set of mathematical equations codifying a "law of nature." Although the word is generally used by scientists more rigorously than the general public to mean an explanatory framework supported by a large body of observational and experimental evidence, even scientists tend to confine the idea of "theoretical science" to the practice of developing mathematical equations to represent a large body of phenomena. While mathematical, computational, and quantitative approaches have important roles in biology, confining the definition of theory to these efforts fails to capture the texture of theory in biology. In Chapter 2 this report adopts a more flexible description of theory as a "family of models" that can be, among other things, physical, visual, verbal, mathematical, statistical, descriptive, or comparative. The models need not even all be entirely consistent with each other (just as it is sometimes useful to model light as a wave, sometimes as a particle), the important characteristic being that the model is a representation of some aspect of nature for the purpose of study. Using this view of biological theory makes it clear how ubiquitous it is in scientific practice. For example, if one's model of the genome suggests that only protein-coding regions are important for development, one may adopt an RNA extraction technique that selects only transcripts with poly-A tails. An alternative model that includes a functional role for noncoding sequences in development would require a different extraction technique. A scientist whose model of cellular robustness rules out the possibility of life below pH 3 or above 90°C will not look for bacteria in the human stomach or in the hot springs of Yellowstone. Explicit recognition that one's theoretical and conceptual framework is affecting choices throughout one's research—from the tools used, to the experiments done, to the interpretation of the results and more—may help stimulate truly innovative and transformative research.

Because theory in biology sometimes corresponds poorly with common stereotypes of theoretical science, biologists and others often fail to recognize its importance. Yet theory is clearly an integral part of the process of biological research and is vital to its success. It is time for biology to take a step back and think carefully about balancing the attention being paid to theory in relation to observation, experiment, and technology development. Would an explicit emphasis on the theoretical and conceptual component of biological research be fruitful, and if so, how would that best be done?

Facilitating Future Advances in Biology: Achieving a Balance

The emergence of a new insight is, by its very nature, unpredictable. In retrospect, however, it is possible in many cases to dissect the relative contribution of theory to the great discoveries of the past. But is it also possible to look at biological research today and determine whether emphasis on one area or another would be most likely to drive innovation and transformation of the field? The topic of transformative research was recently the focus of a National Science Board report, *Enhancing Support for Transformative Research at the National Science Foundation* (May 7, 2007). That report states that "[t]ransformative research is defined as research driven by ideas that have the potential to radically change our understanding of an important existing scientific or engineering concept or leading to the creation of a new paradigm or field of science or engineering. Such research is also characterized by its challenge to current understanding or its pathway to new frontiers." This study's Statement of Task asked the committee to consider whether biology might benefit from an intensive focus on developing theoretical or conceptual foundations: in other words, to consider whether transformative moments would be likely to be driven by a focus on theory.

The increasing fragmentation of the practice of biological research into subdisciplines makes it challenging for biologists to recognize theories that cut across biological scales. The body of knowledge about biological systems has grown so vast so rapidly, and the variety of approaches is now so numerous, that it has become impossible for any one scientist to stay fully abreast of the cutting edge of research—where experiment and observation are actively generating new theories and models (and vice versa)—throughout the full range of biological research. Perhaps even more challenging is the effort to understand enough about other scientific disciplines to know whether the research being done on a specific biological question could inform, advance, or build on research being done outside biology.

Key Questions

The committee chose to illustrate the role theory can play in answering broad questions in the field of biology and addressing grand challenges for society by developing a set of questions that have relevance across many subdisciplines of biology. These questions consider those characteristics that are unique to living systems and are questions that perhaps only the study of living things can answer.

The questions vary. Some focus on characteristics that are similar across many biological scales, while others focus on the incredible diversity of life. Still others take an explicitly theoretical point of view. The committee makes no claim that this set of questions is comprehensive, but simply aims to give a set of important examples of how explicit attention to theory might contribute to answering these kinds of questions—questions that would be difficult to address through a traditional approach. The goal was to choose questions that would illustrate the many connections across biological scales and subdisciplines, not to cover the field comprehensively, nor to identify which new areas of research are the most important or promising. Inevitably, this approach precluded covering any area in depth and made it impossible to include all of the many interesting and innovative areas of current biological research.

The questions are listed below. The summary at the beginning of the report gives a brief overview of each question, and within the body of the report a separate chapter addresses each one.

1. Are there still new life forms to be discovered?
2. What role does life play in the metabolism of planet Earth?
3. How do cells really work?
4. What are the engineering principles of life?
5. What is the information that defines and sustains life?
6. What determines how organisms behave in their worlds?
7. How much can we tell about the past—and predict about the future—by studying life on earth today?

Technology

The focus of this report is on the current and future roles of theory in biology, but it is clear that technological progress will continue to play a critical role in biology research and that it will continue to contribute thereby to advances in theory. Biology has been transformed dramatically in the past decade by technology for the measurement and observation of biological systems and their parts. There are three key features of this technological transformation: (1) the digital information in the genomes of organisms can be fully known; (2) measurements of molecular constituents of cells and their interactions (proteins, gene expression into mRNA, metabolites, molecular complexes) can be global; and (3) these measurements can be dynamic so that time-dependent changes can be seen on a whole-system scale. One of the effects this has had on some areas of biology is that networks can begin to be inferred, dynamic models built, and hypotheses formed based on global dynamic data. This emphasizes the potential for building computational models that are much more useful for explanation and for prediction than ever before. It is likely that a major part of biological research—including the development and testing of models and theories—in the future may be done *in silico*. This report hopes to avoid the stereotype that theoretical science is, at heart, a computational and mathematical exercise: Computation is blind and mathematical modeling is pointless without experimental verification and the development of fundamental concepts and frameworks. Nevertheless, advances in our ability to digitize, store, manipulate, compare, look for patterns in, and interconnect different kinds of biological information represent a technological advance that contributes to all areas of biological practice, from observation, to experiment, and to hypothesis testing, as well as the elaboration of theory.

Understanding the Elephant

In its deliberations on the role of theory in biology, the committee was reminded of the old tale, with roots in African, Indian, and Chinese folklore, of the blind men and the elephant. The tale is told in the following poem, written by a contemporary of Darwin's and published in 1878, just a few years before Darwin's death.

Elephant illustration © Jason Hunt (naturalchild.org/jason)

It was six men of Indostan,
To learning much inclined,
Who went to see the Elephant
(Though all of them were blind),
That each by observation
Might satisfy his mind.

The *First* approach'd the Elephant,
And happening to fall
Against his broad and sturdy side,
At once began to bawl:
"God bless me! but the Elephant
Is very like a wall!"

The *Second*, feeling of the tusk,
Cried, "Ho! what have we here
So very round and smooth and sharp?
To me 'tis mighty clear,
This wonder of an Elephant
Is very like a spear!"

The *Third* approach'd the animal,
And happening to take
The squirming trunk within his hands,
Thus boldly up and spake:
"I see," -quoth he- "the Elephant
Is very like a snake!"

The *Fourth* reached out an eager hand,
And felt about the knee:
"What most this wondrous beast is like
Is mighty plain," -quoth he,-
"'Tis clear enough the Elephant
Is very like a tree!"

The *Fifth*, who chanced to touch the ear,
Said- "E'en the blindest man
Can tell what this resembles most;
Deny the fact who can,
This marvel of an Elephant
Is very like a fan!"

The *Sixth* no sooner had begun
About the beast to grope,
Then, seizing on the swinging tail
That fell within his scope,
"I see," -quoth he,- "the Elephant
Is very like a rope!"

And so these men of Indostan
Disputed loud and long,
Each in his own opinion
Exceeding stiff and strong,
Though each was partly in the right,
And all were in the wrong!

MORAL,
So, oft in theologic wars
The disputants, I ween,
Rail on in utter ignorance
Of what each other mean;
And prate about an Elephant
Not one of them has seen!
                —John Godfrey Saxe (1816-1887)

Each biologist interprets biological phenomena using the data and the tools at hand and a theoretical framework, often acquired through years of education and practice. Molecular biologists seek to explain the elephant by exploring the workings of its genome, ecologists by determining the elephant's role in its environment, neuroscientists by figuring out how the elephant senses and reacts to that environment. Developmental biologists look at how the elephant develops from a single fertilized egg, and evolutionary biologists seek the path by which the elephant came to be the way it is. All combine theories, experiments, observations, and inferences to understand something about the elephant. Unlike the blind men, all are well aware that the elephant cannot be explained by its genes, environment, or history alone. Also, use of this metaphor should not be taken to mean that the committee believes that all biologists should be working at the level of the "whole elephant." Detailed research (the "reductionist" approach) will continue to be critically important and productive. Nevertheless, answers to such questions as "Why is the elephant so large?," "How will global warming affect the elephant?," "How many elephants are needed to preserve the species from extinction?," and "What would be the consequences of extinction?" clearly require input from all areas of biology. Combining insights from different scales and explicitly linking them to see how different approaches complement each other, and to see larger patterns, will allow a richer conceptual basis for "understanding the elephant" to be built. By explicitly giving theory equal status with the other aspects of biology, biological science can become even more productive in the 21st century.

2

The Integral Role of Theory in Biology

He who loves practice without theory is like the sailor who boards ship without a rudder and compass and never knows where he may cast.

—Leonardo da Vinci
(http://www.brainyquote.com/quotes/authors/l
/leonardo_da_vinci.html)

This chapter describes several different ideas about scientific theories, emphasizes the diversity of theoretical activities throughout biology, and discusses ways in which theory is integral to each specific kind of scientific activity, including experimentation, observation, exploration, description, and technology development as well as hypothesis testing. Biologists use a theoretical and conceptual framework to inform the entire scientific process, and they frequently advance theory even when their work is not explicitly recognized as theoretical. Explicit recognition of the many entry points of theory into the scientific enterprise may provide greater opportunity for developing new concepts, principles, theories, and perspectives in biology that would not only enhance current scientific practices but also facilitate the exploration of cross-cutting questions that are difficult to address by traditional means.

THEORY AS PART OF THE PROCESS OF SCIENCE

A good scientific experiment, like a good story, has a beginning, a middle, and an end (Galison, 1987). It is satisfying to describe the scientific method as a linear narrative beginning with hypotheses to be tested and then proceeding to experimental design, execution (funding, equipment and material procurement, set-up and manipulations, measurement and data collection, compilation of results), evaluation of evidence, and formulation of new hypotheses. In the occasional blockbuster scientific story, this process culminates in the emergence of a transformative new insight into nature—the recognition of the cell as the basic unit of life, of mitochondria and chloroplasts as evidence of past symbioses, of plants' ability to turn CO_2 and sunlight into O_2 and sugars. This is rarely the way it happens, however. Real empirical practices turn out to be a good deal more complicated and a good deal less linear. The traditional story of scientific method leaves as a mystery the important question "Where do new hypotheses come from?" But like a bad television screenplay, the mystery is dissipated by focusing the plot elsewhere, on the problem of confirming or falsifying hypotheses—the logic of justification—rather than the psychology of discovery (Popper, 1959).

Each of the steps in this narrative is treated as a black box, when in fact both historical contingency and scientific judgment (in other words, the theoretical and conceptual framework within which the scientists are operating) are at work throughout the narrative, connecting the testing of hypotheses with the generation of new theory. For example, the technologies, protocols, and instruments that are chosen as means of experimentation also appear to have "life cycles." Their endings or disappearance, like experimental methods in the broad sense, can come from anything from a change of interest, to new discoveries that render them obsolete, to new inventions or procedures that replace them. Decisions to use new instruments, to carry out experiments in new ways, or to take notice of odd or puzzling results do not come out of nowhere but instead are informed by the scientists' theoretical framework. The ways in which experimental approaches evolve again hints at more complexity than the standard plot allows.

Scientific observation is likewise complex, although it is often thought of as no more than merely "looking." To count as observation in science, "looking" usually requires a sophisticated approach, involving instruments and elaborate protocols embedded in technical practices that frame and shape both the observations and the reports of the results (Hacking, 1983). The things scientists want to observe are rarely easy to see, hear, taste, smell, or touch unaided by instruments or concepts. The things biologists want to observe are not only complex in their own rights but are embedded in complex structures or communities. Indeed, merely choosing what to observe—and how—is, in fact, profoundly affected by the theoretical and conceptual framework of the observer. Scientific observation is, in other words, as much a matter of thinking in the right way as of looking in the right direction. "The early bird gets the worm," first and foremost because she had the idea to get up early to see if the worms might be more plentiful then. Indeed, observation is fully as active and interventionist as experiment and, in the right context, observation can *be* experimental because the essence of experiment is not manipulation but rather comparative judgment (Bernard, 1865).

Experiment, technology development, and observation all seem to be clearly and familiarly embedded in complex social and technical practices involving people with varied skills, interests, and backgrounds and can appear to be divorced from theory. Theory seems to be different and abstract, the *product* of purely conceptual work to formalize empirical knowledge achieved by science, rather than a living part of the material practice and process of science. Indeed, theory is often described in *opposition* to practice. The word "theory" can be used to describe many different things. It can mean an *idea* behind a hypothesis or the status quo to be challenged; a speculative glimmer of an idea before anyone has tested it; or a well-confirmed, authoritative idea that expresses nature's laws and provides explanations, unification, and means of control after a community of experimenters, observers, and technologists have done their work—but it is infrequently seen as an integral component of each step of the scientific process. Despite this common impression that science is a process and theory its product, however, theory does not merely describe, codify, and enshrine scientific knowledge. It does all of that, and much more, but it cannot be easily dissected out from the body of the scientific enterprise. The many uses of the word "theory," in science as well as in popular culture, not only suggest that theory involves a rich set of practices and processes but also reflect the complexity and variety of theoretical work in science and its value to society more broadly.

A TALE OF TWO THEORIES

The word "theory" serves so many purposes in the English language that confusion is almost inevitable. While anyone who has taken a high school science course has been taught that the word "theory," when used in science, means more than a hunch or an unproved idea, there is nevertheless the tendency to think that some scientific "theories" are more established than others. For example, theories that include mathematical equations and describe a range of physical phenomena that most people have experienced, such as those describing motion or the behavior of gases, are sometimes seen as rising above the designation "theory" and achiev-

Box 2-1
Stephen Jay Gould on the Theory of Evolution

Well, evolution *is* a theory. It is also a fact. And facts and theories are different things, not rungs in a hierarchy of increasing certainty. Facts are the world's data. Theories are structures of ideas that explain and interpret facts. Facts do not go away when scientists debate rival theories to explain them. Einstein's theory of gravitation replaced Newton's, but apples did not suspend themselves in mid-air, pending the outcome. And humans evolved from apelike ancestors whether they did so by Darwin's proposed mechanism or by some other, yet to be discovered.

Moreover, "fact" does not mean "absolute certainty." The final proofs of logic and mathematics flow deductively from stated premises and achieve certainty only because they are *not* about the empirical world. Evolutionists make no claim for perpetual truth, though creationists often do (and then attack us for a style of argument that they themselves favor). In science, "fact" can only mean "confirmed to such a degree that it would be perverse to withhold provisional assent." I suppose that apples might start to rise tomorrow, but the possibility does not merit equal time in physics classrooms.

Evolutionists have been clear about this distinction between fact and theory from the very beginning, if only because we have always acknowledged how far we are from completely understanding the mechanisms (theory) by which evolution (fact) occurred. Darwin continually emphasized the difference between his two great and separate accomplishments: establishing the fact of evolution, and proposing a theory—natural selection—to explain the mechanism of evolution. He wrote in *The Descent of Man*: "I had two distinct objects in view; firstly, to show that species had not been separately created, and secondly, that natural selection had been the chief agent of change. . . . Hence if I have erred in . . . having exaggerated its [natural selection's] power . . . I have at least, as I hope, done good service in aiding to overthrow the dogma of separate creations."

SOURCE: Gould (1994).

ing the status of "laws." Thus the "theory" of evolution, which describes a process of change that is ubiquitous but less often recognized as part of everyday experience than is the steam from a kettle or the acceleration of an object falling to the ground, is seen to be somehow less demonstrably true or scientific than the "theory" of gravity.

However, from a scientific point of view, the two theories have equivalent goals in the sense that both seek to explain and interpret a set of facts. As Stephen Jay Gould memorably wrote (see Box 2-1), "Facts do not go away when scientists debate rival theories to explain them."

The phrase "evolution is just a theory" reflects this tendency toward invidious comparison with well-established laws of "real" sciences like physics. Such a view of evolution might have been apt in 1838, soon after

Darwin's return to England from his five-year voyage on *HMS Beagle*. At that time, Darwin wrote his private "D Notebook" on transmutation while reflecting on the implications of Malthus's idea that population growth inevitably outstrips food supply, a full 20 years before publishing *On the Origin of Species* (Darwin, 1859). "Just a theory," "a hunch," or "an educated guess" can certainly mark the *beginning* of a theoretical enterprise, which in Darwin's case blossomed in a wealth of investigations from the 1830s to the 1870s, followed by the work of evolutionary biologists for more than a century since his death in 1882. Darwin's core principles of his theory of "descent with modification," that is, his mechanism of evolution by natural selection—variation, fitness, and heritability—were first articulated in his "E Notebook" on November 27, 1838 (Barrett et al., 1987). Together with Malthus's principle of population, they form the conceptual core of a theory as profound, as central to biology, and now as well established as Newton's theory of motion. Scientific and public reactions to Darwin's theory upon its publication in 1859 took it to go "beyond the facts," as was Newton's widely attacked "occult" principle of gravity after its publication in 1687. But evolutionary theory has moved beyond Darwin's early insights, just as physics has moved beyond Newton's. Curiously, Newton's "laws of motion" are no less celebrated (nor less useful) for having turned out false (in the wake of relativity and quantum mechanics), while the scientific credentials of Darwin's theory continue to be doubted despite its continuing success in guiding empirical research in a wide variety of biological sciences. It is interesting that at least some physicists no longer describe physical theories in terms of "laws of nature," noting that even such a "well-tested and well-established understanding of an underlying mechanism or process," as the standard model in physics unifying strong and electroweak interactions among fundamental particles, "can never be proved to be complete and final—that is why we no longer call it a 'law'" (Stanford Linear Accelerator Center, 2007).

Though dismissive claims about major scientific theories still play a role in popular debates about the place of science in society and culture, they have little influence on theory development in the sciences, other than as warnings against rash speculation, hasty generalization, and delusions of grandeur at the beginning of a line of theoretical work. It is necessary to look beyond common usage and popular stereotypes to understand the role of biological theory in contemporary science.

To improve our understanding of the role of theory in biology, the view of theory needs to be expanded beyond the traditional concept of a "law of nature" to one that illustrates how the variety of theoretical practices and modes of representation, explanation, and prediction in biology reflect the complexity and diversity of the phenomena that the theory studies. It is important to have a rich concept of theory and the theoretical enterprise in order to understand the many roles of theory in the advancement of biological science, in facing the grand challenges of 21st-century biology, in evaluating how best the biological sciences can be integrated with other sciences, and to ensure that students are able to comprehend and appreciate the patterns and processes behind the wealth of biological facts that are accumulating at an accelerating pace.

VARIETY OF MEANINGS OF THE WORD "THEORY"

The two extreme definitions of the word "theory"—a speculative idea or a mathematical "law of nature"—both serve poorly as descriptors of the role that theory plays in the science of biology. Equating the word "theory" with "hypothesis" is another source of confusion. More broadly useful is an emerging definition of the word "theory" to mean a family of models. This alternative understanding of the word captures the diverse relationships among theories, laws, hypotheses, and models in modern biology and makes it easier to see that biology is a deeply theoretical enterprise, but not one in which theory is understood in opposition to practice, experiment, or observation or focused narrowly on developing a set of master equations.

Theory as Speculation

The view that theory is untested speculation is often accompanied by the view that once "proved," theories turn into facts. Some think of Darwinian evolutionary theory, for example, as mere speculation on grounds that it hasn't yet *proved*, by experiment or observation, that natural selection has produced new species of organisms. Others judge evolution to be pseudo-science, claiming that it *cannot* provide such proof and that, when properly explored, is found inconsistent with the laws of better theories, such as thermodynamics. "[T]heories do not," however, "turn into facts by the accumulation of evidence" (NRC, 1998, p. 6). Nor should the claim that evolution is a theory (speculative or not) be confused with the claim that evolution is a fact. The *fact* that life is genealogically organized by descent, with modification, from a common ancestor should not be confused with the *theory* that the pattern of diversification of life is primarily due to natural selection. Statements about nature state *facts* if they are true, regardless of whether humans have proved them to be so or not. As has been seen, however, it is not at all obvious that successful scientific theories, such as Newton's, must be *true* in order to succeed and be useful. If Newton's theory is false, then it does not state the facts, at least not in the way popular culture demands. The idea that Newton's theory is "approximately true," even while literally false, requires a different account of theories than the traditional one in which successful theories state the true laws, or facts, of nature.

Theory as Quantitative Laws

The idea that mathematical expression is the hallmark of genuine theoretical sciences, while others are simply "less mature," takes physics as a gold standard to which other sciences must aspire, even though it is not obvious that the aim and structure of successful physical theories are well suited to the phenomena of biology or the social sciences.

Examples of important qualitative theories in biology include the circulation theory of the vascular system, the cell theory of living organization, theories of ecological succession, the impact theory of the extinction of dinosaurs, and the theory of evolution by natural selection. Whether qualitative theories such as these win silver or bronze rather than the gold of quantitative theories like Newton's or Einstein's is a matter for debate. Nor is it always clear whether mathematical expression of biological theories would better serve science than their qualitative forerunners. The best mathematical biology is strongly driven by clear concepts. The old joke about the theoretical biologist who began a lecture with the words "Consider a spherical cow . . ." exploits the general lack of understanding of the entry point of mathematical theory. It may or may not be sensible to consider a sphere as a first approximation for the shape of a cow. If the question concerns the phylogenetic relationship of the Bovinae, then the sphere approximation would be laughable, but if the question concerns a calculation of the worldwide release of methane gas due to bovine digestion, then perhaps a spherical approximation might be sensible.

Theory as Hypothesis

Scientists sometimes use the word "theory" as a synonym for "hypothesis" to mean a claim about nature that is intended for empirical testing. Scientists generally recognize that theories and hypotheses can be well or poorly supported by evidence (facts) and that they must sometimes work with weakly supported theories or hypotheses for lack of something better. A "working hypothesis" is commonplace in science. A theory doesn't cease to be a theory because it is confirmed, and a bad theory doesn't cease to be scientific just because it is falsified. More importantly, scientists are well aware of many of the idealizing assumptions they need to make in order to understand, explain, and predict nature and that this means they expect their ideas to be literally false, even if explanatorily productive (Cartwright, 1983; Wimsatt, 1987). Moreover, science is always in process, so scientists can expect theories, hypotheses, and evidence to change over time with continued investigation. A static theory is a dead theory, one that no longer drives research. The popular view of theories as final, conclusive, finished, backward-looking codifications of scientific knowledge is inconsistent with the equating of theory and forward-looking hypotheses, since the former are expected to capture the most durable parts of scientific knowledge— laws of nature—while the latter may well fail testing in the next experiment or observation.

It is important to clarify the difference between hypotheses and theories. In the traditional understanding of science, one starts with a theoretical framework for the particular system of interest. This framework then provides the starting point for a hypothesis (sometimes an innovative or imaginative or inspired hypothesis) that seeks to explain or predict the behavior of the system of interest. The next step is to observe the system or to perform an experiment. The resulting data are then used to confirm or disconfirm the hypothesis (and perhaps the initial theory). When hypothesis and data agree, the theory is confirmed; when not, the theory is disconfirmed. Theories guide the construction of hypotheses for testing but are not themselves put at risk of falsification by a single observation or experiment. Understanding the role of theories in biology should include the broad organizing function of theories to coordinate and direct whole research programs and provide the basis for explaining broad patterns of empirical phenomena.

THEORY AS FAMILIES OF MODELS

The limitations of treating biological theories as candidates for universal laws of nature, or grand empirical hypotheses, or even untested speculations can be addressed by adopting a different viewpoint: that theories are collections or "families" of models. A scientific model is a representation of some aspect of nature for a purpose of study (Levins, 1966, 1968; Giere, 1988; Lloyd, 1988; Teller, 2001; Wimsatt, 2007). Most biological systems are too complex to be described by a single model; a family of related models is more appropriate. Modes of representation in models are quite diverse, including verbal, mathematical, visual, and physical. Darwin used words to present evolutionary models, while Robert May used mathematics to formulate ecological models of deterministic chaos. Many molecular biologists and neurobiologists use diagrams to depict causal structure in their models, for example, of how transcription factors regulate gene expression or how neurons interact in brain circuits. Prior to computers, chemists often built elaborate physical models of molecular structures (for more information on modes of model representation, see de Chadarevian and Hopwood, 2004). Models serve as representations because modelers intend them to. This relativity to scientists' purposes means that models represent nature only in relevant respects to limited degrees of accuracy (Giere, 1988, 1999; Teller, 2001). Watson and Crick intended their original wire and metal models of DNA to represent its helical structure in terms of bond angles among the constituent kinds of metal pieces representing the geometrical structure of groups of atoms (purine and pyrimidine bases), but they did not intend to represent the color of atoms as metal gray, the backbone as a continuous, homogenous wirelike strand or made of metal, or the distance between base pairs as several inches (see Giere et al., 2006).

Whatever the mode of representation of a particular model, mathematics will frequently be involved in the scientific process of explaining, predicting, or controlling nature. If not in the formulation of the theoretical model itself (or the integration of a family of mathematical models to express a general law), mathematics will be involved in the expression of predictions from the model (as in the use of equations to predict the temperature at which particular DNA sequences will melt into separate strands), or in the aggregation of observations and measurements into useful data sets (as in the population sciences and increasingly in global databases in the molecular sciences), or in statistical procedures to evaluate the test of a hypothesis, or in the design and operation of instruments and computer simulations. A diagram might represent the causal path in a biological mechanism, for example, of the impact of predators on prey in population ecology, or the distribution of characters in a phylogenetic tree, or from a neural circuit to a particular behavioral output. To understand the dynamic *operation* of such causes, mathematical representations are usually necessary and often mathematics is needed to build a visual representation from data in a database. Increasingly, videography is used to capture dynamic aspects of natural phenomena visually and animation can be used to display dynamic aspects of structural models. At a minimum, mathematical tools are needed to develop and use these visual display technologies, since most are computer based, and to depict empirical data stored in databases. Of all the skills required to do biology, mathematical and computer skills may require the most focused and sustained attention by the K-12 and university education systems and in the continuing education of successful scientists. Quantitative approaches are a critical link between theory and other biological practices.

The traditional view of theories, built around the reductionist ideal of the most powerful explanations emanating from the lowest levels, anticipates a single, general, realistic, and precise formal representation in a master equation for a given domain (or even for all of science). There is actually no single best, all-purpose model for any natural phenomenon (Levins, 1968). There can be several, even incompatible, models of the same phenomenon because each can represent separate aspects and our purposes may be quite varied. Teller (2001) points out that physicists sometimes model water as an incompressible continuous fluid medium and at other times as a collection of discrete particles. Biologists sometimes model organisms in a population as genetically homogeneous but ecologically variable or, conversely, as genetically variable but ecologically homogeneous (Roughgarden, 1979). Practical purposes and interests typically force scientists into tradeoffs in their models among virtues of accuracy, precision, realism, and generality as well as fruitfulness in stimulating new ideas, testability of hypotheses, and intelligibility of concepts.

Accepting such tradeoffs is not a sign of theoretical weakness or limitation, so long as empirical results are tested for robustness to the idealizing assumptions of any given model. Acknowledging tradeoffs, in other words, does not mean that the science is somehow bogus, but rather that the "conceptual engineering" that goes into model building and robust analysis of results is an important and explicit part of the theoretical enterprise (Wimsatt, 2007). Quantitative predictions of the precise abundances of organisms in a model of an ecological community with an unrealistically low number of interacting species might trade off (for reasons of analytical tractability or computational power) against qualitative predictions of increase or decrease with a more realistic number of community members, for example. Computer simulation may bridge that particular tradeoff (facilitating numerical solutions to analytically unsolvable equations and quantitative predictions about many species), but other idealizations in computer programs may limit generality in other respects (e.g., that every simulated member of a given species is assumed to be genetically identical). Computer models of interacting molecular networks that are being developed to understand gene regulation represent a spectrum of approximation methods: from binary state, to Boolean models, to systems of differential equations, to stochastic random models of molecular interactions, and hybrids of all these types. Simplifications are key features of all these models. Levins (1968) conjectured that at most one could maximize two out of three desirable features a model could have: generality, realism, and precision. His point was that our pragmatic interests in biological phenomena, together with our limited ability to work with and understand complex representations, suggest that we may never reach the dream of a "final theory" and that, more importantly, we need to evaluate the conceptual tradeoffs carefully and with much thought if we are not to be led into error. These issues will come up in attempts to construct computational models of the cell that include more and more molecular species, their concentrations, properties, and interactions.

Anything *can* serve as a model for anything else, but whether a model is *useful* in a particular context depends on the respects and degrees of relevant similarity between a model and what it is intended to represent (Teller, 2001). A fruit fly may (or may not) be a useful genetic model for a human, while a Buick may rarely be a useful physical model of a black hole, but either can count as a model, given some specified sense of relevant similarity and some specified or implied degree of accuracy that can guide evaluation of the "fit" of a model to the world. Mathematical models play an especially useful role in most sciences because of their special role in rigorously formulating assumptions and establishing formally the consequences of their operation. Although even very simple mathematical models can exhibit extremely complex behavior—for example, chaos in simple growth models in population ecology or neural network models in cognitive science—often the rigor of mathematical analysis or, increasingly, the power of computation and simulation to extend calculation and reasoning abilities (Humphreys, 2004) is needed to trace clearly the implications of assumptions that cannot be easily interpreted or understood either intuitively or verbally.

One virtue of understanding theories as families of models rather than as laws of nature is that models need not be expressed in mathematics nor even in statements, though language and mathematics are two key ways humans have to communicate relevant similarities. One concrete object (e.g., styrofoam balls on sticks) can represent another (the solar system, a molecular structure). Biologists often talk about "animal models" for diseases or for physiological processes. And laboratory systems of organisms exposed to various conditions have often been taken to serve as models for particular biological processes, such as the flour beetle system (Tribolium species) as a model for ecological competition (Park, 1941; see Griesemer and Wade, 1988) or fruit fly systems (Drosophila species) as models for evolutionary, gene transmission, behavioral, or developmental processes. In other cases, a particular phenomenon serves as a model for thinking about and constructing others, as when a particular set of molecular interactions in the promoter region of a gene are studied and used as a basis for exploring genetic regulatory systems in other cases or more generally (see Keller, 2000).

Another particularly useful aspect of recognizing biological theory as families of models is that it sheds light on the very fruitful practice of comparing models. In many situations, for example, formal mathematical models can be crucial in helping investigators determine when their qualitative models actually are adequate. Biologists often come up with "word models" about processes which then are shown to be inadequate when one tries to actually implement a formal mathematical model or construct a computer algorithm. "When things get too complicated for human intuition and language, scientists turn to math and models" (von Dassow and Meir, 2004, p. 245). Building formal mathematical models and running simulations is a tool of experimental work that can be useful as one method for testing the adequacy of our understanding and for understanding how interactions among components can give rise to system behavior. As the increasing ease of collecting large amounts of data makes it more and more possible to study system-level interactions, mathematical and computational models are becoming increasingly important to many areas of biology.

Quantitative approaches, from formal mathematical models, to simulations, to pattern recognition algorithms, have another very important value: By requiring logical discipline and a formal methodology, they can be a powerful tool in hypothesis development and prediction. In some instances, large data sets can themselves serve as experimental resources. One can argue that the field of molecular biology, for example, "has finally inverted the habit of biological inquiry. Instead of using phenomenology and perturbation experiments to deduce some mechanism, and then uncovering facts one by one to support that hypothesis, modern biologists increasingly turn to large-scale exploration (e.g., DNA microarrays, genome sequencing) to generate a mass of facts whose relevance is eventually established by phenomenology and from which mechanistic understanding might hopefully emerge" (von Dassow and Meir, 2004, p. 245). Large-scale methods vary considerably in their ability to deliver reliable quantitative data. DNA sequencing is highly reliable, while large-scale gene expression data are only semiquantitative and most large-scale interaction maps from yeast two-hybrid assays and other methods are not even reproducible from lab to lab. Dynamical mathematical and computer models are some tools for coping with these ever-growing masses of data, and computational methods can often be used to improve the usefulness of data of variable quality. Importantly, not all of these methods demand mathematically tractable models. Computers can enable researchers to test hypotheses without having to come up with master equations. Monte Carlo simulations, for example, can test thousands of complicated scenarios and provide a different kind of demonstration of the "robustness" of a hypothesis than would a mathematical model. Just as biologists' theoretical and conceptual frameworks drive their choice of experimental and observational strategies, theory will play a critical role in making the best possible use of large data sets. Indeed, the ability to test hypotheses computationally (experimentation *in silico*) may be one of the most important future sources of theoretical breakthroughs in biology. The accumulation of biological data and its storage, maintenance, and accessibility are challenges today. Theoretical approaches to data analysis are likely to be highly productive but will require scientists, or collaborative teams, that combine biological expertise (both theoretical and experimental) with computational and mathematical competence.

CROSS-CUTTING QUESTIONS

In the course of subsequent chapters, it will be made clear that there are many theories, concepts, and principles that operate at the many levels of organization that biologists now study, on timescales from the picoseconds (10^{-12} s) of vibrational state changes of biomolecules to the 4.5 billion year history (10^{17} s) of planet Earth, and on size scales from elementary particles such as the electrons (10^{-15} m diameter) that are exchanged in biochemical reactions to the planet itself (10^7 m diameter), the physical characteristics of whose surface and atmosphere have been profoundly affected by life, from the evolution of oxygen-generating life forms billions of years ago to anthropogenic climate change today.

A model-based view of scientific theories complements the traditional view of (correct) scientific theories as sets of (true) statements of laws of nature, enriching our understanding of the theoretical enterprise and its multiple roles in empirical biology. If there are universal laws of nature, they are as likely to be discovered through study of a variety of models as by a direct search for them. The production of a variety of models to explore a given biological phenomenon from different perspectives creates opportunities, and deep need, for renewed attention to theory and support for theorists willing to question basic assumptions and standard approaches. Support for theoretical work in science, because of theory's many entry points into biological practice, may require investment in both low-risk traditional as well as high-risk radically transformative approaches, since the robustness of empirical results to the idealizing assumptions of conventional models cannot properly be evaluated without worthy alternatives to compare. This report frames a series of questions about life that cut across established disciplinary perspectives while drawing on shared principles or theories that are central to all biological subdisciplines, including basic principles of evolution (life is descended from a common ancestor and natural selection is a key mechanism of change), of cell biology (all life is made of cells), and of heredity (specific evolved mechanisms of intergenerational information transfer account for genealogical relationships).

3

Are There Still New Life Forms to Be Discovered?
The Diversity of Life—Why It Exists and Why It's Important

In an age when people can visit the bottom of the ocean or the inside of a volcano from the comfort of their living rooms, it may seem strange to ask whether there are any new life forms to be discovered. But, in fact, the extent of life's diversity has not yet been determined. Just 30 years ago, scientists on board the deep-sea submersible *Alvin* discovered an unexpectedly diverse community of sea life in hydrothermal springs 2.5 kilometers below the surface of the ocean near the Galapagos. *Alvin*'s crew found a diverse community, including giant tubeworms, huge clams, and ghost-like crabs thriving around the hot submarine springs (Van Dover, 2000). This complex ecosystem was fueled not by the harvesting of the sun's energy by photosynthesis but by energy derived by bacteria from the hydrogen sulfide spewing from the vents.

The study of life's diversity involves more than just going into the world or the laboratory and looking for new things. The places we look, the tools we use, and the experiments we do are influenced by our theoretical and conceptual understanding of the limits of life, the mechanisms of evolution, and the role and significance of diversity. Conversely, new observations and experimental results are constantly forcing us to adjust our theoretical framework. This chapter gives examples of the extent of diversity at several different scales in biology and illustrates the many roles that theory plays in the study of these different kinds of diversity.

The fantastic creatures that populate the ocean's hydrothermal vents are just one example of situations where discoveries have triggered an expansion of biology's theoretical framework. Our views of where life can exist have been regularly revisited; organisms are being discovered in habitats—from the human stomach to more than a mile underground—where conditions were thought to be too harsh to allow life. New birds, plants, and mammals are still found with some regularity. Entomologists name and describe new insect species at a rate of about 1,500 per year. The evidence that some genes have been conserved throughout evolution and the availability of polymerase chain reaction to survey those genes made it possible to begin exploring the diversity of the microscopic world. Suddenly, tiny organisms that appeared under the microscope to have only a few basic and uncomplicated body forms were revealed to be unimaginably diverse—in fact a new kingdom of life, the Archaea, was discovered to be as different from bacteria as bacteria are from eukaryotes (Woese et al., 1990). The advent of high-throughput sequencing and sophisticated computational analysis has allowed biologists to begin to plumb the diversity of the microbial world, and it appears that life at the microscopic level is vastly more diverse than biologists ever imagined. A recent survey of microbes in the ocean using an approach called metagenomics not only revealed thousands of previously unseen genes but hundreds of novel protein families. Families of proteins that were already known, like the rhodopsins that absorb light in the human retina, were found to have hundreds of distinct members in the ocean sample (Bejà et al., 2000, 2001). The vast numbers of new genes are not necessarily mere variations on known themes; the potential functional diversity—in other words, proteins and synthetic pathways that carry out currently unknown reactions—to be found in microbial communities is enormous (e.g., Venter et al., 2004; Zhang et al., 2006; Gill et al., 2006).

What is the significance of discovering one more beetle, one more bacterium, or one more protein? One answer lies in the incredible diversity of functions that evolution has generated. Nature has foreshadowed our technical developments, and functional biodiversity can be a fertile source of ideas for technology. For example, a group of neuroscientists has found a parasitic fly that can locate the sounds of its hosts—field crickets—with unparalleled accuracy. Remarkably, the fly's ears are tiny and only one-half millimeter apart (Mason et al., 2001). The fly's ears have inspired the design of directional microphones and a new generation of directional hearing aids. Another example is a group of brittlestars (relatives of sea stars) that have turned their skeletons into a visual system made up of arrays of microscopic lenses (Aizenberg et al., 2001). The lenses detect light and allow the animals to find dark hiding places on the ocean bottom. Such small lenses are beyond current human engineering capability. However, their precisely curved shape and the way they are arrayed are prompting engineers to create novel optical devices.

Recognizing that nature provides a vast toolbox is only one motivation for studying life's diversity. The complex interconnected web of living species is critical to human life. Humans depend on the living world in countless ways. The connection between biological diversity and the stability of ecosystems is only imperfectly understood. Clearly, the living world will continue to evolve in response to environmental change, but from the human perspective the time scale of that adaptation is crucial. Understanding the role of biological diversity and how it is generated, maintained, and lost is a critical goal for 21st-century biology.

MAKING SENSE OF LIFE'S DIVERSITY

The Diversity of Species

The effort to identify, describe, and name distinct organisms in a systematic and coherent framework has been underway for hundreds of years. These activities are called taxonomy. Currently, systematists—a name change that reflects a change in the underlying conceptual basis of classifying diversity—study the details of organisms' characteristics and the interrelationship of characteristics between different organisms (e.g., whether the middle finger of a human corresponds to the middle digit of a bird; Wagner and Gauthier, 1999). Systematists use such comparisons to organize organisms into a classification system that rationally groups similar organisms together. Both the methods by which these activities are carried out and the description of the astonishing diversity of organisms are works in progress. They are essential works, for a system of nomenclature and classification is necessary in order to organize knowledge about the millions of species, known and yet to be described. Clearly a system of classification requires the underpinning of a robust theoretical framework.

The still commonly taught hierarchical Linnaean form of classification (species, genus, family, etc.) was proposed and developed by Carl Linnaeus (1707-1778) a century before *The Origin of Species*. While Linnaeus is credited with devising a system for the orderly classification of species, in fact, his own classification schema for plants grouped them strictly according to the number and arrangement of their reproductive parts, leading to groupings, like castor beans with conifers, that now sound illogical. Linnaeus's binomial naming system has survived, but subsequent taxonomists followed the example of naturalists like John Ray (1628-1707), who had begun to classify organisms on the basis of groups of morphological and physiological characteristics. The prevailing theory underlying the study of diversity at that time was that there existed a fixed number of species and that the job of naturalists was to name and catalog each of them in a logical way. The fastidious work of specimen collection followed by comparative morphology and physiology, while carried out within what is now seen to be a false theoretical framework (that the number of species was fixed and that species did not change over time), nevertheless provided the body of data that Darwin used to develop the new theory of descent with modification. With the addition of Darwin's theory of evolution, comparative morphology and physiology became a richer undertaking, and it became possible also to integrate extinct life forms into the tree of life by studying the characteristics of fossils. The classification systems developed by comparative taxonomists from John Ray forward, indeed, correspond surprisingly well with the genetic data that began to emerge after the identification of DNA as the molecule of heredity. The theoretical relationships between organisms proposed by taxonomists can now often be demonstrated through computational comparison of their genetic sequences, a field known as "phylogenetics." Indeed, the theoretical hypothesis of descent with modification provided a rich source of potential experiments that could be carried out bio-informatically. The comparison of gene sequences through phylogenetics has confirmed that many of the taxa (hierarchical groups of organisms such as "arthropods" or "insects") recognized by pregenetic classification schemes correspond to evolutionary lineages. The theoretical basis of modern systematics rests on grouping species into taxa, or "clades," that, according to the best interpretation of data, have descended from a common ancestor and thus form one branch of the great tree of life, the phylogeny[1] of all organisms.

Classification of organisms into named grouping entities (i.e., taxa) is a nontrivial task, but there has been enormous progress in phylogenetic systematics, owing both to the development of increasingly sophisticated statistical methods and algorithms for inferring phylogeny and to DNA sequencing. In particular, DNA sequences provide data that can be treated quantitatively and are more broadly comparable across the diversity of life than the type of data that predated the molecular revolution (Kim, 2001a). Indeed, the recent availability of genome-scale information and whole genomes enhances our ability to construct phylogenetic relationships by considering multiple related genes, genomic rearrangements, genomic content, or even functional relationships of genomic components (e.g., Boore, 2006; Wolfe and Li, 2003).

Phylogenetic descriptions of diversity are immensely useful, partly because they capture a great deal of information and partly because they give us a guide to the history of organisms and their characteristics. Phylogenies summarize a great deal of history and can be used for tracing the evolution of the traits and molecular characteristics of even extinct organisms (see Box 3-1). Historical trends as revealed by phylogeny can have important applications as well. Just as the knowledge of past trajectory is used to gauge the future landing site of a thrown football, phylogenetic reconstruc-

---

[1]A phylogeny is a tree-like diagram where branches represent evolutionary lineages and leaves represent current organisms.

Box 3-1
What Could Dinosaurs See?

By comparing current DNA sequences, biologists can deduce the sequences of those genes in the ancestors of current species. Chang (2003) and colleagues investigated the characteristics of the visual pigments (rhodopsins) of archosaurs, the ancestors of dinosaurs, birds, and crocodiles. Phylogenetic analyses allowed the comparison of rhodopsin genes of a wide variety of living organisms and generation of the best estimate of what the gene sequence would have been in their distant, common ancestor. Most interestingly, the theoretically deduced gene sequence could be cloned into laboratory bacteria where it was shown to code for a functional protein. The function of the reconstructed protein could then be tested. It was shown to be most sensitive to light of the wavelength of 508 nm—a slightly longer wavelength than that perceived by modern vertebrates—suggesting that archosaurs may have been able to see in dim light. Thus the work both sheds light on the lifestyles of extinct organisms and validates the general approach of theoretical estimation of ancestral gene sequences, followed by direct laboratory study of the reconstructed proteins.

———————————

SOURCE: Chang et al. (2002).

tions can be used to estimate prospective evolution of rapidly evolving organisms such as influenza viruses or antibiotic-resistant bacteria and hence to develop vaccine and treatment strategies (Smith et al., 2004; Koelle et al., 2006).

There remain both practical and conceptual limitations to using phylogenetic trees to create classifications. The limitations fall into two basic categories. First, the mathematics is extremely complicated. Second, while evolution is driven by general rules of natural selection, there is also an element of chance. Many possible genotypes may have the same phenotype and fitness, so that the eventual descendant whose sequence is studied today could have many equally possible ancestors.

There are biological, statistical, and computational challenges in phylogenetic reconstruction. First, on the biological side, there are approximately 1.5 million described organisms and vastly more undescribed organisms. It is still a huge challenge to obtain phylogenetically relevant information from such a large collection of organisms—the development of new technologies such as massively parallel sequencing will be critical to solving this problem. Accurate estimates of phylogeny require statistical models of evolution as a base starting point. There are still

considerable problems in constructing biologically reasonable yet computationally approachable statistical models. For example, it is very difficult
to resolve the branching order among lineages that diverged either very
recently or very long ago. Solutions to these problems will likely require
statistical models of molecular processes other than simple single base-pair
DNA mutations as employed now.

The major challenge of phylogenetic tree estimation lies in the computational domain. Consider that for only 10 species there are over 34 million
possible alternative phylogenetic trees, and for 30 species there are more
numbers of possible trees than there are atoms in the universe! The goal of
phylogenetic estimation algorithms is to select the optimal tree among such
impossibly large numbers of possibilities. The magnitude of this computational challenge has led a computer scientist to exclaim "There are enough
problems, already formulated or yet to be developed, to keep teams of algorithm designers busy for many years, and just the right combination of real
data, credible simulation, and scaling issues to make *phylogenetics* [italics
ours] the ideal testing ground for algorithm engineering" (Moret, 2005). In
other words, the problems of phylogenetics are challenging enough to test
the mettle of the state-of-the-art approaches of mathematicians, engineers,
and computer scientists. The importance of getting it right, however, is high
because the tree of life is our map to life's history and to the relationships
among organisms. The tree of life is used as a guide for research and to
find out the origin of traits, including why human bodies are vulnerable to
certain kinds of failure. The seemingly inexplicable narrowness of our birth
canal and the persistence of genes that cause diseases have their origin in
our evolutionary history, and why humans live as long as we do can be better understood when scientists find our position in the tree of life and trace
how the working features of organisms have evolved along its branches
(Nesse et al., 2006).

The Challenge of Microbial Diversity

A basic concept underlying phylogeny is that diversity arises from the
branching of lineages from a common ancestor rather than from fusion
(hybridization) of distinct lineages. The many species of finches on the Galapagos arose from a single ancestor species whose descendants specialized on
different food sources, not from mixing and matching between an ancestral
finch and other specialized birds. Therefore, evolutionary theory suggests
that evolution should create genealogical trees rather than networks. This
idea captures the broad pattern of evolution and has been immensely useful,
yet it can be problematic for some organisms, especially the noneukaryotes.
Early results of metagenomics studies (see Box 3-2) demonstrate that the
genomes of bacteria and archaea are extremely variable. Organisms that

---

Box 3-2
Theoretical Questions That Can Be Addressed with Metagenomics

One of the most exciting recent developments in microbiology is community genomics or metagenomics. Instead of trying to isolate and study individual microbial species, practitioners of this approach characterize DNA from entire mixed microbial communities. The metagenome of a habitat includes the genomes of all the microbes living in that habitat. Thus, in metagenomics, genes and their functions are studied independently of the species from which the DNA is derived. Metagenomics makes accessible the diversity of the microbial world and has considerable potential to transform biologists' view of life. A recent report (*The New Science of Metagenomics*, NRC, 2007) expanded on the conceptual and theoretical questions that may gain new answers in the light of metagenomic research.

"Decades of genetic, molecular, and biochemical dissection of microbial life have revealed the detailed structure and inner workings of several bacteria and archaea. Although there is much more to learn even about model organisms, such as *E. coli,* many individual pathways for nutrient cycling, gene regulation, and reproduction are understood at a satisfying level of precision. But these processes in the majority of microbes remain unknown and knowledge of the evolution and ecology of microbial communities lags far behind cellular microbiology. Basic ideas that organize biologists' understanding of the living world may need refinement in the face of greater understanding of community function.

What is a genome? The number of genes in the genome of a free-living bacterium ranges from 500 to 10,000 or more; the largest bacterial genomes are more than twice the size of the smallest eukaryotic genomes. In contrast, the genomes of many parasitic or symbiotic microbes are highly reduced, with not nearly enough genes to support them independently of their hosts. As more data accumulate, the definition of what constitutes a microbial genome will be better informed and underlying principles governing genomic plasticity in microbes may emerge. . . . If having a more flexible and dynamic genome structure is a fundamental life-strategy difference between bacteria and archaea, on the one hand, and eukaryotes, on the other, what are its advantages and limits? Can understanding the phenomenon help to explain the emergence of multicellular organisms that have more fixed genomes?

What is the role of microbes in maintaining the health of their hosts? Closely associated microbial communities appear to be a common, if not universal, feature of the physiology of multicellular organisms. These communities contribute to a variety of functions, from digestion to defense against pathogens. All plants and animals, including humans, can be considered superorganisms composed of many

---

would be considered the same species on the basis of the similarity of certain highly conserved genes may be found to have only 50 percent of the rest of their genes in common, with many other genes that are not found in every individual. Microbiologists are developing the concept of a "pan-genome" to describe the set of genes that are shared by all members of a species—animal, bacterial, archaeal, and viral. Using the human as an example, the human "metagenome" might be considered an amalgamation of the genes contained in the *Homo sapiens* genome and in the microbial communities that colonize the body inside and out. The organisms within these communities are collectively known as the human "microbiome." The metagenome of these communities encodes physiological traits that humans have not had to evolve, including the ability to harvest nutrients and energy from food that would otherwise be lost because humans lack the necessary digestive enzymes.

Metagenomics will enable us to address a number of fundamental question. . . . Is there an identifiable core microbiome shared by all humans? How is each individual's microbiome selected? What is the role of host genotype? Should differences in each individual's microbiome be viewed, with the immune and nervous systems, as features of our biology that are profoundly affected by individual environmental exposures? How is the human microbiome evolving (within and between individuals) over different time scales as a function of changing diets, lifestyle, and biosphere? How can this knowledge be used to manipulate microbial communities to optimize their performance in a person or in a population? Most obviously, how does the microbiome affect health, and vice versa? In the future, previously unrecognized microbial involvement with disease states will be uncovered. Many host physiological states with primary genetic or biochemical causation will affect the microbiome in ways that may aid in diagnosis. Of course, these questions do not apply only to humans—study of host-associated microbial communities will contribute to understanding of the physiology of all organisms.

What ecological and evolutionary role do viruses play? Viruses are important not only as pathogens but as agents of lateral gene transfer and catalysts that generate tremendous genetic variation in their specific hosts. Viral activity also has important consequences for turnover of the elements, for example, in carbon cycling in aquatic systems. It has only recently been recognized that virus particle numbers are enormous, often exceeding those of co-occurring cellular life. For example, seawater contains 10 times more bacteriophage than cellular microbes. Estimates suggest the biosphere harbors perhaps as many as 10^{31} viral particles (Edwards and Rohwer, 2005). Given these vast numbers, the influence of viruses on biodiversity and evolutionary catalysis, and their role in biogeochemical cycling, there is considerable interest in characterizing naturally occurring virus populations. Metagenomics has recently provided an important avenue for exploring these ubiquitous and biologically important entities."

SOURCE: NRC (2007).

microbial species (Tettelin et al., 2005). The great variability of microbial genomes is the result of horizontal gene transfer; bacteria and archaea can exchange genetic material by a number of different mechanisms, even with organisms that are distantly related. The prevalence of horizontal gene transfer means that the phylogenetic relationships of microbes may look more like networks than trees and all of an organism's different genes may not have the same phylogenetic relationships. For some microbiologists, the very concept of "species" seems problematic for organisms whose genomes can be so variable, but others maintain that the concept of species will be useful for categorizing noneukaryotic organisms. Until more is known about the extent and pattern of horizontal gene transfer, this conceptual issue will remain open. Horizontal gene transfer is most common in non-eukaryotes, but there is evidence of transfer of genes between symbiotic partners (Hoffmeister and Martin, 2003). While such events may be rare and not affect the overall shape of the tree of life, their existence provides evidence of additional sources of genetic variability on which natural selection can act. Defining the role of horizontal gene transfer is only one of several fundamental theoretical issues raised by the study of microbial communities (Box 3-2).

Genetic Diversity Is Itself Diverse

Biological diversity is more than species diversity. The study of biodiversity usually focuses on changes in species numbers in time and space. Life, however, is diverse at all scales. There is diversity in the organization of genomes; in genes and their protein products; in genetic networks and the molecular machines they assemble and regulate; in strategies for defense against pathogens, mobility, and detection and reaction to the environment; and in the morphological, behavioral, and physiological characteristics of individuals within species. At all these levels, there is constant interaction between the theories currently used to describe the extent and consequences of diversity and the relentless flow of new examples of diversity.

Genome Size

The genome of an average mammal has around 3 billion pairs of nucleotides. This is about a hundred times longer than all the letters in a 20-volume encyclopedia arranged in a line (Avise, 2004). Genome sizes vary from a few thousand base pairs in viruses to 600,000 base pairs in some bacteria to more than 200 billion base pairs in some animals. Genome sizes do not correlate with position on the tree of life—bacterial genomes range from 0.6 Mbp (*Mycoplasma genitalium*, an intracellular pathogen) to approximately 1 Mbp for many free-living bacteria, to 10Mbps for the filamentous cyanobacterium *Nostoc punctiforme*. Invertebrate genome sizes vary by more than three orders of magnitude, from 29Mbp (the root-knot nematode) to 63 billion bps (an amphipod), while vertebrates vary about 400-fold in size (from the 342 Mbp of the green pufferfish to the 129 billion base pairs of the marbled lungfish), as indicated in Figure 3-1. The lack

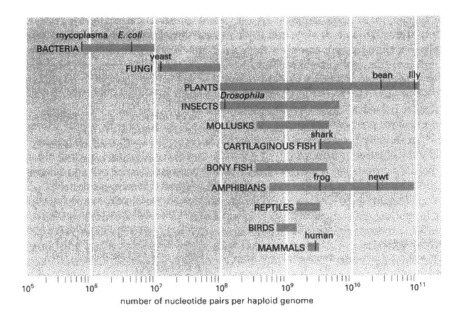

FIGURE 3-1 Genome size in various organisms.
SOURCE: Molecular Biology of the Cell, 2002, by Alberts et al. Reproduced with permission of Li and Sinauer and Garland Science/Taylor & Francis LLC.

of obvious correlation between genome sizes, phylogenetic relationships, or organism complexity has stimulated the development of a new area of biological inquiry and experiment. The sheer size of the genome can accommodate a lot of variation, and indeed genomes can differ enormously even within a single species. Stephens et al. (2001) have estimated that random pairs of homologous DNA sequences from humans would differ in about 1 out of every 1,000 base pairs, meaning that one human differs from another at an average of 3 million sites. Individual base pairs are not the only place at which genomes can vary; a recent study of 270 individuals found that approximately 12 percent of the genome showed differences in gene copy number from one individual to another (Redon et al., 2006). Repetitive genetic elements and transposable genetic elements (segments of DNA that can move from one spot to another in the genomes of their hosts) may be found in different places in different individuals.

Neutral Theory

Variation in the genetic code is the raw material of natural selection and thus evolution. However, it is only relatively recently that it has been understood how vast the extent of genetic variation is, how many different forms it can take, and how its magnitude can be estimated. Levels of genetic variation within a population are determined by important natural processes, including mutation, demographic structure and fluctuations, and natural selection. Genetic variation across species is governed by similar factors, albeit at a longer time scale. Thus, understanding the extent and limits of variation is a critical component of a theoretical understanding of evolution. Prior to the molecular era, the magnitude of genetic variation was controversial. One camp (the "classic" camp) argued that genetic variability was low and that most individuals in a population shared the same form of each gene. The alternative camp (the "balance" camp) maintained that variation was high and that most individuals had different forms of the same gene (Lewontin, 1974). The controversy simmered for years because genetic variation was so difficult to measure. The history of the exploration of genetic diversity is a good example of how scientific progress comes about from the interaction of the development of new technologies, the data generated, and the theory developed to make sense of the data.

Only about 40 years ago, in 1966, several laboratories used the newly developed method of gel electrophoresis to separate the proteins produced by a gene. The method suggested that the genomes of humans and fruit flies had a lot more variation than anybody expected. The broad applicability of the initial observations was debated, but the spread of the measurement technologies soon revealed that the larger-than-expected variation was common for many different genes. Thus, the protein electrophoresis era helped to resolve the theoretical debate about estimates of genetic variability and shifted the debate from the amount of genetic variability to its causes. The "balance" school argued that genetic variation was the outcome of natural selection (in the jargon of population genetics, of superiority of heterozygotes, frequency-dependent selection, and variation in fitness among habitats). In the meantime, the development of methods to sequence proteins produced data that suggested that in vertebrates new amino acid variants become fixed in a typical 100 amino acid protein at the rate of about 1 per 28 million years. Extrapolating to the size of the typical genome, Motoo Kimura in 1968 made calculations to show that such a rate would imply one amino acid variant being replaced in the entire population once every three years. If such a replacement were due to new advantageous variants, then all individuals without the variant must be eliminated from the population—an unsustainable "substitution load" for the population. Thus, he boldly hypothesized that most of the genetic variation in a population must be neutral variants that are randomly changing in their frequency. Neutral mutations are genotypic changes that do not cause phenotypic changes. Because natural selection is conspicuously absent from the equations, the theory is called "neutral." Initially, the idea that mutational variants would have neutral effect on selective dynamics was controversial—mutation should surely be either advantageous or deleterious. However, current molecular data conclusively confirm that dominant molecular variation in populations must be neutral. The theory of natural selection is not thereby overturned—it is clear that many DNA sequences have experienced natural selection. The addition of the "neutral" theory of genetic change, however, makes us look at variation in the genome in a new way, because other changes in the genome are the result of neutral processes (Bustamante et al., 2005). The theory of neutral evolution is a canonical example of theory providing an explanatory framework for new data with far-reaching implications for understanding the process of evolution and the functional consequences of molecular changes.

The evidence that some differences between genomes have functional significance and some are neutral naturally led to efforts to find ways to distinguish between the two. Theoreticians have developed methods to detect the telltale molecular evidence of natural selection and thus to quantify the relative importance of selection and neutral processes. Relevant examples have emerged from the fine details of the major histocompatibility complex system (Schaschi et al., 2006), from the self-incompatibility mating system in plants (Charlesworth et al., 2005), and from the evolution of the mechanisms that plants use to resist the attacks of their natural enemies (Rausher, 2001) (see Box 3-3).

The sequence of bases is not the only information stored in the genome; chemical modifications of DNA, such as methylation, and the three-dimensional packaging of DNA have important effects on when various genes are expressed. These "epigenetic" mechanisms (mechanisms that affect the expression of genes or inheritance of traits in ways other than changing the sequence of the DNA) are yet another example of common phenomena that have a role in the origin, maintenance, and loss of diversity.

Gene Duplication

Sometimes a genome contains multiple copies of related genes. These gene families originated by gene duplications. In all three domains of life, a large proportion of all distinct genes were generated by gene duplication (Zhang, 2003; Bowers et al., 2003). Estimates of the percentage of duplicate genes range from 17 percent in some bacteria (Himmelreich et al., 1996) to 65 percent in *Arabidoposis thaliana* (*Arabidopsis* Genome Initiative, 2000). Duplicated genes can be grouped in families that share

---

Box 3-3
Plant Resistance to Pathogens

In plants, detection of a pathogen infection initiates a cascade of processes beginning with the death of cells at the site of infection and activation of a systemic protection system that attacks the pathogen. A receptor protein that recognizes the invading pathogen initiates the defense cascade. The specificity of the receptor to the pathogen is one line of evidence for the idea that the receptor is adapted against a pathogen. Several other lines of molecular evidence support this conjecture: (1) The genes for these receptors have undergone large numbers of changes that have led to numerous amino acid substitutions over a short period of time. (2) The rate of base substitutions that lead to amino acid changes (nonsynonymous substitutions) is higher in these genes than the rate of substitutions that do not lead to amino acid changes (synonymous substitutions). (3) The changes in the receptor gene are concentrated in the region that interacts with the pathogen's molecule that elicits the response. (4) Finally, in some of these resistant genes, the phylogenetic evidence suggests that several forms of the same gene are often maintained in the population for a very long time, presumably as a result of natural selection that favors the maintenance of several different variants ("balancing selection"; Stahl et al., 1999). The availability of extensive sequence data has led theoreticians to develop an arsenal of statistical techniques to deduce the probable action of natural selection on DNA sequences (Ford, 2002).

---

common ancestors and in which the members can have diverse functions, but a common theme emerges: What is the fate of a gene after it duplicates? After a gene duplicates in an individual, its fate is similar to that of a new mutational variant. If the duplication is neutral, it has a tiny probability of being fixed. Sometimes the presence of a duplicate gene can be selectively beneficial because two genes make more RNA and protein. In this case, purifying selection acts to maintain the function of the two copies (Wagner, 2002). Sometimes the duplicated gene is redundant and the accumulation of deleterious mutations in one of the two genes transforms it into a pseudogene (a nonfunctional copy of an active gene). This process seems to be one of the sources of the many pseudogenes in genomes. Harrison et al. (2002) suggest that there is one pseudogene for every two functional genes in the human genome.

Selection can favor the retention of two or more functional duplicates if the sequences of the two genes diverge and lead to different functions (see Box 3-4). RNase1, for example, has a double function: It is secreted by the pancreas into the intestinal lumen where it digests RNA, and it is expressed in many tissues where it defends against viral infection. Colobine

Box 3-4
New Function Through Gene Duplication

One of the important outcomes of gene duplication is the origin of novel, albeit often related, function. One nice example is the genes that code for the red- and green-sensitive opsins in humans, which were generated by the duplication of a sex-linked gene in hominoids and Old World monkeys, and which give us trichromatic vision. Howler monkeys, a group of New World monkeys, evolved trichromatism independently through a duplication of the same gene in the x chromosome. The olfactory receptor (OR) genes that form the largest gene family in mammalian genomes are another good example. A high percentage of these genes are "pseudogenes" that have lost their function, presumably as a result of disuse. Interestingly, the frequency of pseudogenes among the OR family members differs greatly among species. While humans, nonhuman primates, and mice have roughly the same number of OR genes (about 1,000), in humans about 60 percent of these are pseudogenes, while nonhuman apes have about 30 percent, and mice have only about 20 percent (Menashe et al., 2003). What are the factors that may cause this large interspecific variation in the proportion of pseudogenes in the OR family?

Gilad et al. (2004) randomly sequenced 100 distinct OR genes from each of 18 primate species—four apes, six Old World monkeys, seven New World monkeys, and one prosimian. They found that Old World monkeys had roughly the same percentage of OR pseudogenes as nonhuman apes (about 30 percent) but a much higher percentage than New World monkeys (about 17 percent), except for howler monkeys. The percentage of OR pseudogenes in the howler monkey was about 30 percent, much closer to that seen in the Old World monkeys and apes than in its New World relatives. The higher frequency of pseudogenes in the OR family must have evolved independently in howler monkeys and Old World monkeys. Recall that howler monkeys share trichromatic color vision with apes and Old World monkeys. The evolution of trichromatism seems to have coincided with the deterioration of the sense of smell. This leaves the question of why humans have such high frequencies of OR pseudogenes. Gilad et al. (2003) speculated that cooking food reduces the need to identify odorous toxins in food, which may be denatured by heating. Paradoxically, cooking, which we associate with delicious aromas, may have diminished our capacity to smell diverse odors.

monkeys have two copies of the RNase gene (RNase1 and RNAse1B), one of which retains the presumably ancestral function (RNAse1) and another that helps the monkeys digest bacterial RNA (RNAse1B). Unlike other primates, colobine monkeys are foregut fermenters and must digest large amounts of DNA from the rapidly growing fermenting bacteria in their guts (Zhang, 2003). Zhang et al. (2002) found that since duplication, RNAse1B had much higher rates of nucleotide substitutions at nonsynonymous sites

(sites where a DNA base change results in the incorporation of a different amino acid in the protein) than at synonymous and noncoding sites, providing evidence of selection for a function that complements that of the ancestral gene.

Diversity in Functional Noncoding Sequences

The "central dogma," which states that the role of DNA is to code for RNA, which in turn codes for protein, focused scientists' attention on documenting the variation in protein-encoding genes for much of the past 50 years. The prevailing theory suggested that understanding such variation would explain much of life's diversity. This focus prevailed even when biologists studying complex multicellular organisms, such as mammals and plants, knew that only a small percentage of their genomes actually encoded proteins. The rest of the genome was often referred to as "junk" DNA, which was thought to be made up of mostly remnants of transposable elements, DNA that selfishly existed only to replicate—with little impact on genome function—or pseudogenes. As more data accumulate, it is becoming clear that at least some of this "junk" DNA does contribute to functional diversity and thus could contribute to variation upon which selection can act. The diversity found in the portion of the genome previously considered more or less inert is vast when one compares sequences between closely related species; considerable diversity is sometimes found even within a species. How much of this diversity contributes to function is still unknown, but results deriving from comparative genomics and high-throughput methods to examine genome-wide expression patterns combined with functional genetic analyses in fungi, plants, and animals challenge our previous conceptions and suggest much remains to be learned about how genome diversity dictates functional diversity. The discussion below is not meant to be comprehensive but serves to illustrate that while the "central dogma" is broadly correct for protein-coding genes, it is apparent that our theoretical framework explaining how genomes function requires expansion.

A large portion of many eukaryotic genomes is made up of repetitive sequences, existing in tens, hundreds, thousands, or millions of copies within a genome (Morgante, 2006; Jurka et al., 2007). It is this repetitive portion of genomes that is usually not conserved at the nucleotide level between even closely related species, although organisms as different as plants and animals do share the same classes of sequences. Some of these sequences are simple tandem repeats (Armour, 2006), stretches of DNA where the same short sequence is repeated hundreds or thousands of times. The number of repeats can vary so much between individuals that these sequences are excellent markers for genetic and forensic studies (Armour,

2006). While the function of most of these sequences is unknown, there are a number of diseases associated with variations in triplet repeat lengths (Mirkin, 2006).

Other repetitive sequences are derived from reverse transcription of RNA molecules into DNA, with subsequent integration into the genome. There are several different examples of such sequences, called transposable elements, which can move around the genome. Much of the repetitive DNA in the genome seems to consist of defective copies of these transposable elements that have suffered mutations, so they can no longer transpose. Some of these sequences have expanded tremendously with a single type of element contributing millions of copies to a genome. Functional transposable elements (those that encode the proteins necessary for transcription, reverse transcription, and integration and thus can still move) are found in much lower numbers (Ding et al., 2006). These active elements cannot only move themselves, they can move related defective elements and reverse transcribe mRNA or structural RNAs to generate pseudogenes. Still other classes of transposable elements transpose via DNA replication mechanisms (Morgante, 2006; Jurka et al., 2007). Integration of any of these types of sequences can affect the expression of adjacent genes through the regulatory sequences they contain or disruption of regulatory sequences at the insertion site. Thus, diversity in where these sequences are located within an individual's genome can have consequences for gene and genome function. In some species, such as humans, the insertion of many of these defective transposable element sequences was ancient. However, there are subsets of elements that have moved more recently and insertion sites for these differ from person to person. In a species such as maize, many more transposons are currently active relative to what has been described in mammals, which likely contributes to the amazing diversity between inbred lines in terms of numbers and organization of genes and gene fragments (Morgante, 2006).

It is commonly thought that gene fragments are the ultimate in junk DNA; it is hard to imagine a function for a fragment of a gene inserted into a noncoding region between genes. However, the observation that many of these sequences are transcribed, sometimes on both strands, combined with the discovery of a number of RNA-mediated gene-silencing mechanisms involving double-stranded RNA, raises the possibility that in some instances these gene fragments contribute to diversity of gene expression patterns by targeting functional genes containing the same sequence. RNA interference, RNAi, is an evolutionarily conserved mechanism in fungi, plants, and animals that generates short 21-23 nucleotide RNAs (siRNA) from double-stranded RNA, which then target corresponding mRNAs for cleavage (Rana, 2007). MicroRNAs (miRNAs) are a class of short RNA that are encoded in the intergenic nonprotein-encoding regions of animal and plant genomes. miRNAs are produced through processing of imperfect RNA hairpins and depending on their degree of complementarity elicit either translational control or mRNA cleavage, resulting in gene silencing that is essential for animal and plant development (Zhang et al., 2007). In addition to post-transcriptional gene silencing, there are RNAi-related pathways that regulate gene expression by modifying DNA methylation or how DNA is packaged with the result that the functional gene sharing the RNAi sequence is silenced (Matzke and Birchler, 2005; Chan et al., 2005; Grewal and Elgin, 2007). The repetitive elements discussed previously are major targets of these RNAi-related transcriptional silencing pathways. When the pathways are disrupted, there can be significant consequences to the organism; other genes can also be regulated via this mechanism (Zaratiegui et al., 2007). That there appear to be so many evolutionarily conserved and regulated sequences and regulatory pathways outside the traditional genes is a relatively new observation, ripe for theoretical input.

In addition to the types of sequences discussed above, comparison of genomes across diverse species from vertebrates, invertebrates, plants, and yeast have identified a large fraction of conserved nonprotein and non-RNA- encoding sequences under selective constraints (e.g., Waterston et al., 2002; Kaplinsky et al., 2002; Siepel et al., 2005). While some of these sequences are likely to be regulatory, transcription factor binding sites do not necessarily show high sequence conservation even though a fraction can be functionally conserved (Dermitzakis and Clark, 2002; Fisher et al., 2006). From studies done to date, it is clear that noncoding DNA sequences can have significant effect on phenotype and are subject to natural selection (reviewed in Bird et al., 2006). However, the functions of most conserved noncoding DNA sequences are unknown, let alone the functions of nonconserved noncoding DNA sequences; it is possible that species- or genera-specific sequences may serve a much wider range of roles than currently imagined.

In summary, the relatively recent recognition of new RNA pathways for controlling gene regulation, as well as the extensive transcription of the human (ENCODE Project Consortium, 2007) and plant genomes (Stolc et al., 2005; Hanada et al., 2007) that results in the majority of DNA sequences being represented by transcripts, combined with the lack of understanding of evolutionary constraints on noncoding DNA, suggest much remains to be learned. A focus on only the variation in the protein-encoding portion of the genome is unlikely to lead to full understanding of life's diversity or the mechanisms and evolution of genome function. New computational methods and new theory will be required to fully understand the function of the vast majority of genomes, the noncoding DNA.

Diversity of Molecular Function

The previous section discussed the many ways that diversity can be generated at the level of genes and genomes. Over the billions of years of evolution, this variation has produced vast numbers of genes that encode functional proteins. Determining the function of a protein in one organism can be useful for predicting its homolog's function in another organism. However, even for organisms that are very well studied, like yeast or humans, the functions of all gene products are not yet known. Determining the function of each gene product experimentally is not only inefficient but can also be misleading as the activity of a protein may differ according to context. Therefore, improving our ability to predict computationally the function of gene products, or to understand the functional consequences of mutation, is an important challenge.

Since the mid-1990s, the increasing availability of genomic sequences and molecular diversity data has stimulated interest in the fields of bioinformatics and computational biology. The recent discovery of great molecular diversity in functional genomic elements other than protein-coding sequences (e.g., ENCODE, 2007) as described above suggests even greater theoretical challenges in this area. Accurate computational prediction of molecular function from sequence information and the use of comparative diversity data in genomic annotations remain a great challenge. Genomic sequence information includes both coding and noncoding functional sequences. Function prediction from this information includes everything from prediction of molecular structure from protein sequences and RNA sequences to organization of these structures into functionally predictive frameworks. What ultimately is required is a collection of models that allow us to construct a "map" from sequence to molecular function to organismal function. Conceptually, this requires first a construction of a "sequence feature space"—that is, a distillation of sequence features relevant to function prediction and a relational metric (distance measure) using those features (Kim, 2001b). At present, most standard approaches involve statistical characterization of known examples—the expanding information on molecular diversity greatly helps these approaches. However, the ultimate goal, especially when presented with entirely novel sequences from, for example, metagenomics projects (see Box 3-2) where even the organism of origin is unknown, is the derivation from first principles of a functional theory of biomolecular sequences. At present, determining protein function from gene sequence is hard. It is complicated by the fact that proteins are part of complex machines, and many years of work may be required to determine the full set of interactions and functions of any protein. However, once it has been done, scientists can benefit from the ability to extrapolate across the phylogenetic tree to other organisms. It is clear that a systematic computational/theoretical framework for the prediction of function would provide a critical boost in efficiency compared to empirically driven, eclectic approaches.

Diversity of Social and Behavioral Systems

As if life were not diverse enough at the molecular, genomic, species, functional, and community levels, organisms also have wildly diverse behavioral and social interactions. Even a brief survey of the range of diversity at this level would be difficult, so this section discusses one particular topic that crosses genetic, evolutionary, behavioral, and social boundaries: the area of sex, gender, and sexuality. This particular area is controversial and often even politically charged, but incontrovertibly reproduction is an essential characteristic of all living organisms. The debate over whether the accepted theoretical framework regarding the role of sexual selection in evolution, initially outlined by Darwin and subsequently built on for over a century, can accommodate new data and perspectives, serves as an example of the integral and often unacknowledged role of theory in biological research.

Some biologists have drawn attention to many examples of expressions of sex, gender, and sexuality throughout the animal kingdom that are unanticipated by and challenging to the prevailing theoretical framework. Within evolutionary biology, the conceptual treatment of sex roles originated with Darwin's theory of sexual selection. Darwin introduced this theory because of traits like the peacock's tail that are termed ornaments and that are not readily understood as adaptations for survival. Instead, Darwin hypothesized that such traits find their evolutionary value in how they promote mating. The process that causes traits to evolve because of how they contribute to mating is called "sexual selection," which Darwin contrasted with "natural selection," the process causing traits to evolve that promote survival.

When Darwin proposed his theory of sexual selection, he took the peacock and peahen, and the stag and doe, as emblematic of males and females generally. He asserted generalizations like, "Males of almost all animals have stronger passions than females" and "the female . . . with the rarest of exceptions is less eager than the male . . . she is coy" (Darwin, 1871). Darwin amassed examples to support these claims of universality. Sexual selection thus enunciates a *norm* of natural sexual conduct. Species that depart from the sexual selection templates of passionate male and coy female are then seen as "exceptions" meriting special discussion to account for their deviant behavior.

However, there are many species in which males and females are virtually indistinguishable, as with the guinea pigs many people raise as pets, or birds like penguins, where sexes can only be distinguished by careful inspection of the genitals. In other species, males are not passionate, nor females coy, and the females consistently pursue the males. Female alpine accentors from the central Pyrénées of France, for example, solicit males for mating every 8.5 minutes during the breeding season. Ninety-three percent of all solicitations are initiated by the female approaching the male, with the other 7 percent by him approaching her (Davies et al., 1996). This frequent sexual contact greatly exceeds that needed specifically to fertilize the relatively few eggs that are reared.

Or what can be concluded from the seahorse and pipefish, in which the male is drab and the female ornamented, and in which the male raises the young in a pouch into which the female deposits eggs? Such species exhibit what biologists call "sex role reversal." The females are said to compete for access to males, with the males choosing females for their ornaments, resulting in showy females and drab males, the reverse of the peacock. Such a situation contradicts the traditional assumption that the cheapness of sperm invites passionate male promiscuity and the expensiveness of eggs necessitates female coyness during their careful choice of good gene-bearing males. But male seahorses make tiny sperm just as male peacocks do, and female seahorses make large eggs just as peahens do; nonetheless, male seahorses care for the young and female seahorses entrust their eggs to a male's pouch.

In many species, multiple types of males and females, each with distinct identifying characteristics, carry out special roles at the nest both before and after mating takes place. In the sandpiper-like European ruff, black-collared males build nests in small defended territories called courts within a communal display area called a lek. Meanwhile, white-collared males accompany females while the females feed. The white-collared males then leave the company of the females and fly to the lek where they are solicited by the black-collared males to join them in their courts. When the females eventually arrive at the lek to lay eggs, they are romanced by pairs of males—one black-collared male paired with one white-collared male in some courts, as well as by single black-collared males in courts by themselves. Evidently, females prefer to lay eggs in nests hosted by a pair of black-collared and white-collared males at which both males serve as parents, rather than in nests hosted solely by one black-collared male, perhaps because the white-collared male has formed a bond with the females while he was accompanying them during their feeding. Perhaps white-collared males serve as "brokers" who introduce females to the black-collared males, who have not previously had the opportunity to meet females while they were busy setting up and defending courts in the leking area. There are, in fact, many examples of family organizations consisting of trios such as the ruffs, or of species with reproductive social groups that consist of more than one male and one female tending offspring together after mating takes place, or even participating jointly in courtship before mating takes place.

Same-sex sexuality is also evident in many species. In more than 300 species of vertebrates, same-sex sexuality has been documented in the primary peer-reviewed scientific literature as a natural component of the social system (Bagemihl, 1999). Examples include species of reptiles like lizards, birds like the pukeko of New Zealand and European oystercatcher, and mammals like giraffes, elephants, dolphins, whales, sheep, monkeys, and one of our closest relatives, the bonobo chimpanzee.

For some biologists, this cornucopia of diversity in gender expression and sexuality severely strains Darwin's sexual selection theory. At the same time, the last 50 years have witnessed a great expansion of Darwin's sexual selection narrative that was originally focused rather narrowly on secondary sexual characters like peacock tails and deer antlers. Many, perhaps even most, evolutionary biologists do not feel that the accumulation of counterexamples and exceptions has risen to the level of requiring a major overhaul of sexual selection theory. Others argue that, just as the fossil record undermined the theory that each species was individually created and unchanging, these "exceptions" cannot be reconciled with current theory. It is not the role of this report to resolve that controversy but merely to use it as an example of the more universal process whereby observation, experimentation, and the building and testing of models and hypotheses are intimately affected by one's initial theoretical viewpoint and the evolution of that theoretical viewpoint in response to ongoing research.

Diversity in Context

Diversity at the molecular, functional, and organismal levels is multiplied at the environmental level, where groups of species co-inhabit countless overlapping ecosystems. This is the context in which evolution plays out, where all the different kinds of variation at the genetic level provide, or fail to provide, a selective advantage and where external changes in an environment eventually lead to the adaptation, migration, or extinction of local species. The field of ecology has a long history of theoretical approaches to the understanding and prediction of what governs species diversity in different environments, the role of species diversity in ecosystem stability, and the impact of environmental change.

What Governs the Assembly of Communities?

What is it that determines how many and which species will form an ecosystem? How much of the resulting community is due to chance, to history, or to underlying principles of energy and resource availability? The greater our ability to identify underlying governing principles, the better the predictions of the effects of change. According to the competitive exclusion principle, two or more species that are identical in their use of a limiting resource (such as space or food) cannot coexist indefinitely, and only one of the populations will survive competition; if one is competitively superior, exclusion of the others proceeds all the more quickly. Many mathematically formulated hypotheses have been proposed, and tested to various extents, to explain assemblages or communities of coexisting species. The simplest is "niche partitioning," whereby competing species do not fully overlap in resource use, each having a "refuge" resource of which it is the sole or competitively superior consumer. Any textbook of ecology describes examples that conform to this prediction. Such patterns are ascribable both to evolutionary responses of species to each other and to purely ecological processes of assembly, wherein members of a species pool colonize a location and either form a stable population or not, depending on whether or not they "fit."

Resource partitioning among species is not always evident, especially among organisms such as plankton and terrestrial plants. Among the major factors proposed to maintain diversity are predation and disturbance. A panoply of specialized predators (or parasites), each specific to a different prey species, may hold each prey species at a low enough density to enable other species to persist. For instance, specialized consumers of seeds or seedlings may contribute to maintenance of tree species diversity in forests (Janzen, 1970; Connell, 1971). More generalized predators may likewise maintain diversity by preventing competitively dominant prey species from excluding others, although prey species that are less able to escape predation may be eliminated. Likewise, physical disturbances may open sites for colonization, and species capable of high dispersal (or which lie in wait, as do buried seeds) may persist if they can reproduce before they are excluded by dominant competitors. Such "fugitive" species often characterize early stages in ecological succession. This idea underlies a number of models of patch dynamics, including lottery models in which ecologically equivalent species persist almost indefinitely if enough gaps open at random in a sufficiently large landscape.

Lottery models mark a shift in ecological thinking from equilibrium to nonequilibrium models, the most renowned of which may be MacArthur and Wilson's (1967) model of island biogeography, in which the number of ecologically equivalent species on an island is set by rates of distance-dependent colonization and area-dependent extinction. This is the simplest explanation for the dependence of diversity on area, one of the most abundantly documented of ecological patterns, and postulates that the diversity in a local area (e.g., an island) is not determined solely by local interactions but also by the species diversity and dynamics of a larger region that feeds local diversity by immigration. Ecologists have increasingly accepted that this principle holds for local assemblages in continental sites as well, so landscape-level processes and regional species diversity strongly affect diversity and dynamics at a local level (Ricklefs and Schluter, 1993).

MacArthur and Wilson's model was extended, moreover, to continental biotas and to evolutionary time by Rosenzweig (1975), who modeled species diversity as a consequence of rates of speciation and extinction. Hubbell (2001) has developed this approach to its fullest extent in his "neutral theory of biodiversity," in which the population genetic theory of genetic drift is applied to ecologically equivalent species. Although Hubbell does not deny that species often partition resources and are differentially resistant to predation and disease, his model shows that these processes may not need to be invoked to explain the patterns of diversity in many communities, such as abundance distributions of tropical forest trees.

Why Are Some Communities More Diverse Than Others?

Community ecologists have long felt that a theory of species diversity in communities should be able to explain variation in the number of coexisting species among assemblages in different environments and different parts of the world. The challenge may be epitomized by the latitudinal gradient in species diversity: In most higher taxa of plants and animals, diversity is highest in tropical regions and declines toward both poles. On land, diversity declines from warm, wet environments (such as those that harbor tropical wet forest) toward colder high altitudes and toward more arid regions.

Traditional theory assumed both ecological and evolutionary equilibrium: It would not do to say that cold regions have fewer species because they pose special adaptive challenges, since that simply shifts the question to why cold-adapted clades should not have diversified as much as warm-adapted clades have. As many as 100 hypotheses for these patterns have been distinguished (Willig et al., 2003). Many ecological explanations suggested either that plant communities in warm, wet climates have higher productivity, and that this would support more species, or that tropical regions experience less variable climate, so that more specialized species could evolve and coexist by finely dividing resources among them. However, tropical regions are not more climatically stable (they are often more variable in rainfall than temperate regions), and there is little or no evidence that tropical species are more specialized; for example, herbivorous insects in tropical wet forest appear to be no more host specific than in temperate-zone forests (Novotny et al., 2006). The primary productivity of tropical wet forests may actually be lower than that of high-latitude forests (Huston, 1994), and although high productivity might support higher population densities of animal species and therefore reduce their extinction rate, it is hard to see how it would sustain higher plant diversity. In fact, whether species diversity of plants increases monotonically with productivity or peaks at intermediate productivity is a subject of some controversy (Huston, 1994; Gillman and Wright, 2006).

In contrast, nonequilibrium explanations of the latitudinal diversity gradient, advanced in various forms for decades (e.g., Fischer, 1960), are gaining favor. One class of hypotheses holds that speciation rates are higher in tropical regions. The fossil record of bivalves (Jablonski et al., 2006) and of foraminifera and other planktonic organisms (Buzas et al., 2002; Allen and Gillooly, 2006) supports this hypothesis; in fact, bivalve taxa have originated mostly in the tropics and expanded toward the poles. Why, then, should speciation rates have a latitudinal bias? One possibility is that terrestrial tropical species, living in more constant temperatures, are physiologically intolerant of very different temperatures and are less capable of surviving the temperature stress they would experience in dispersing over mountain ranges (Janzen, 1967). Few data bear on this hypothesis, but those few largely support it (Ghalambor et al., 2006). It has also been suggested that high temperature increases rates of mutation and that this heightens evolutionary rates in general and speciation rates in particular (Allen et al., 2006; Gillman and Wright, 2006). A reported correlation between rates of molecular evolution and speciation (Webster et al., 2003) may support this hypothesis (which parts from the traditional supposition of population geneticists that genetic variation is so plentiful that phenotypic evolution is seldom limited by the rate of origin of adaptive mutations).

A more deeply historical view, rapidly gaining adherents, is that the tropics have more species because most clades originated in tropical environments and have remained mostly restricted to them because of the several factors that cause "niche conservatism" (Brown and Lomolino, 1998; Ricklefs, 2004). Until about 30 million years ago, tropical climates embraced a far greater area than they do now; in fact, the diversity of tree species in tropical, temperate, and boreal biomes is correlated with the area typified by those climates during the geological time (Eocene to Miocene) when most clades evolved (Fine and Ree, 2006). This "tropical conservatism hypothesis" (Wiens and Donoghue, 2004) builds on the strong correlation between species richness and geographic area and articulates in modern terms the older hypothesis that there has been more time for diversification in tropical regions (Stebbins, 1974). Plant genera that are distributed across continents have highly correlated latitudinal distributions (Ricklefs and Latham, 1992), exemplifying the long-sustained niche conservatism that is central to this hypothesis. A phylogenetic analysis showed that hylid frogs originated in the tropics, spread only recently into temperate regions, and display a strong correlation between the species richness of a region and when that region was colonized (Wiens et al., 2006). An almost inescapable conclusion is that patterns of species diversity can be understood best by taking into account evolutionary processes over very long periods of geological time.

There is, perhaps, a profound lesson in this brief summary of efforts to develop and test general theories explaining patterns of species diversity. Many models and computational approaches have been brought to bear on understanding the complex relationships linking a community of species to one another and their physical environment. It now appears that at least part—perhaps a large part—of the explanation lies in history. The increasing availability of genomic sequences and refinement of phylogenetic theory will contribute to the validation of this theory, but if the role of historical chance is significant, there are both practical and philosophical implications. If biodiversity depends on evolutionary processes acting on the available genetic reservoir over geological time scales, the loss of species due to rapid, human-caused environmental change has profound consequences on the stock of genetic possibilities for the future. Philosophically, if biodiversity is largely the consequence of natural selection acting on random genetic events in specific communities and environments over very long time periods, the search for underlying, quantifiable, predictable order in the origin, maintenance, and loss of species is made vastly more difficult.

Loss of Diversity

A population or species becomes extinct when its last member dies. Most ecological analyses of extinction follow either a "small population" paradigm or a "declining population" paradigm (Caughley, 1994). The former focuses on risks of extinction faced by small populations even in favorable environments, owing to stochastic fluctuations (Lande et al., 2003). In addition, some local populations ("sink" populations) cannot maintain a positive rate of increase without immigration from other populations and dwindle if immigration is curtailed. In the declining population paradigm, populations are driven to low numbers by deterministic forces, including abiotic environmental changes (in climate, for example), changes in landscape (especially habitat loss), and changes in the biotic environment. Most extinctions of entire species probably are attributable to these kinds of causes.

Even aside from "mass extinction" events such as the K/T extinction (in which the dinosaurs perished) that has been attributed to a bolide impact, "background" extinctions have occurred throughout evolutionary history and have befallen far more than 99 percent of the species that have ever existed. Clearly a species is a transient thing in this statistical sense. Remarkably little is known about the causes of these extinctions, although certain species characteristics, such as broad geographic range, ecological breadth, and high dispersal capability tend to be correlated with longer persistence times (Jablonski, 1995). Still, the ecological factors that cause extinction, and the organism-level or species-level traits that determine survival versus extinction, are little known. Even the factors that limit geographic ranges along environmental gradients, where local populations cannot persist, are understood for very few species (Parmesan et al., 2005). Some of the most immediate current threats to populations and species, however, are anthropogenic and are fairly obvious: overexploitation (especially of large vertebrates and marine resource species) and habitat destruction. Much of conservation biology focuses on understanding how species can be saved in the face of these threats. Models of population dynamics and of dispersal among subpopulations in increasingly patchy landscapes are important tools in conservation.

Extinct species are those that have not adapted to whatever environmental changes befell them. The population genetic theory of microevolution should, ideally, enable us to predict population survival versus extinction, but doing so will require both significant theoretical advances and far more information than is currently available.

The first question is whether or not the environmental change is one that would be expected to trigger an adaptive response. This can occur only if there is a change in the rank order of the fitness of different genotypes. Some changes, however, reduce population size without altering relative fitness. If a critical resource such as food or habitat dwindles, individuals may experience the same resource environment as when it is abundant, so there may be no change in relative fitness. Williams (1966) described such species as "running out of niche" but remaining well adapted to that niche to the bitter end. We need a better understanding of what environmental changes do not alter the regime of natural selection.

When an environmental change does engender selection for adaptive change, there begins a race between a demographic process of declining population size and the evolutionary process of adaptation (Holt and Gomulkiewicz, 2004). The simplest models of adaptation to changing environments envisioned selection on a single quantitative character such as body size, in which the population mean can track a moving optimum, although lagging behind it, and the population can maintain positive population growth if the genetic variance of the character is high enough (Lynch and Lande, 1993). Since directional selection will exhaust initial genetic variation, long-continued evolution will then depend on a sufficiently high rate of mutational input of new genetic variation, which depends on population size. More realistic models must take into account the reduction in population size that results from the lag, the various genetic architectures that a trait may have, and the realistic expectation that the environmental change may impose selection on multiple traits. Population genetic theory has shown that adaptation is likely to be slower, the greater the number of independent characters, or "dimensions" of genetic variation (Wagner, 1988; Orr, 2000), and that genetic correlations among characters may enhance or retard the rate of evolution, depending on where the new phenotypic optimum lies, relative to the multidimensional axis of greatest variation (Lande, 1979; Kirkpatrick and Lofsvold, 1992).

Predicting which species will survive and which will become extinct as a result of an environmental change is an important and exceedingly difficult challenge. Consider the global temperature change, already underway, that inevitably will transpire at a rate that has perhaps never been equaled in evolutionary history (Parmesan, 2006). What aspects of a species' environment will change, what characteristics might, by evolving, provide adaptation to these alterations, and what levels of selectable genetic variation might enable adaptive change in these features are all major unknowns. The negative impacts on populations are not at all limited to thermal stress; they are already known to include phenological (seasonal) mismatch between a species' life cycle and the phenology of its food supply, critical changes in its physical environment (e.g., polar bears depend on dwindling ice floes for hunting seals), and changes in the community of species with which a species interacts (Parmesan, 2006). For any particular species, it would be hard to identify all the characteristics that might be directionally selected, given such a multiplicity of possible impacts. And there is increasing evidence that populations may have little or no genetic variation in some ecologically critical characteristics (Blows and Hoffmann, 2005), such as dessication resistance in flies (Hoffmann et al., 2003), the capacity of herbivorous insects to adapt to certain plants (Futuyma et al., 1995), and the ability of plants to adapt to toxic soils (Bradshaw, 1991). It is perhaps no wonder, then, that species display niche conservatism (Wiens and Graham, 2005) and that the response of most species to Pleistocene glacial/interglacial oscillations was not adaptation to the climatic changes visited upon their original locations but massive, repeated shifts in geographic range as species tracked the climatic "envelope" to which they were already adapted (Williams et al., 2004).

Because of complex ecological linkages, species do not become extinct independently, and the extinction of key species can have cascading effects. For example, overexploitation of fish populations has had devastating effects on coral reefs, kelp beds, and even the pelagic food web (Scheffer et al., 2005). Consequently, ecologists are increasingly concerned that the loss of species diversity may have drastic effects on ecosystem "services" such as productivity and may result in ecosystem collapse. Preliminary models, as well as data on the consequences of marine biodiversity loss, give credence to these fears (see Figure 3-2; Dobson et al., 2006; Worm et al., 2006). The

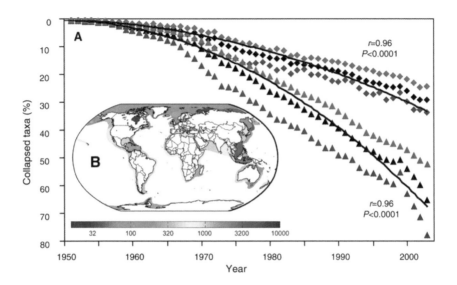

FIGURE 3-2 Global loss of species from large marine ecosystems (LMEs).
(A) Trajectories of collapsed fish and invertebrate taxa over the past 50 years (diamonds, collapses by year; triangles, cumulative collapses). Data are shown for all (black), species-poor (<500 species, blue), and species-rich (>500 species, red) LMEs. Regression lines are best-fit power models corrected for temporal autocorrelation. (B) Map of all 64 LMEs, color-coded according to their total fish species richness.
SOURCE: Worm, B. 2006. Impacts of Biodiversity Loss on Ocean Ecosystem Services. *Science* 314:787-790. Reprinted with permission from AAAS.

possibility of devastating ecological effects of human impacts underscores the need for increasing theoretical and empirical studies of the interplay between species diversity and ecosystem characteristics.

Extinction is, then, one of the least well-understood phenomena in ecology and evolutionary history. In evolutionary biology, a deeper understanding is required of the causes of niche conservatism, the dimensionality of genetic variation, the factors that determine variability (the capacity of characters to vary), and the nature of and linkages between genetic and demographic processes in changing environments. Theoretical and empirical advances are needed in ecology to address questions about the abiotic and biotic factors that can extinguish populations and about the linkages among species and ecosystem processes that might accelerate losses in diversity, productivity, and ecosystem health.

CONCLUSION

The diversity of biological systems extends from the molecular to the global scale and all of the levels are linked. Survival or extinction of a species and the stability of an ecosystem may depend on the level of random, neutral genetic variations that have built up in individual members of various species over time and on the balance between the size of those species' populations and the rapidity of change in their environment. At all levels, general theories to explain and predict diversity would be a great advance: from defining the evolutionary relationship of species, to predicting the function of proteins from gene sequence, to relating the form and functions of organisms to their genomes, to predicting the stability of ecosystems from their constituent species. The vastness of the diversity and the important, but as yet undefined, role of chance and history in biological systems make the development of such theories a grand challenge indeed.

4

What Role Does Life Play in the Metabolism of Planet Earth?

For the first 3 billion years of Earth's history, all life was confined to the ocean and was entirely microbial. That life was electric. The earliest forms of life evolved a set of mechanisms for extracting energy from the chemicals around them and from sunlight by transferring electrons from one element or molecule to another. All life on Earth continues to rely on this ability to move electrons, and the electron transfer reactions, invented by the earliest bacteria and passed on to all other living organisms, form a nested set of biologically catalyzed elemental cycles. The elemental cycles are coupled to geochemical and geophysical processes that, in concert, have sustained life on Earth from the start of the geological record about 3.8 billion years ago. The biological conversion of solar energy into chemical energy ultimately became the primary source of energy for all life on the Earth's surface. Through an obscure series of evolutionary occurrences, the highest energy state that evolved produced oxygen as a byproduct of splitting water; the hydrogen atoms were used to form organic matter from carbon dioxide. The energy in this "primary" organic matter is used by other organisms for energy and growth. Furthermore, the resulting accumulation of oxygen in Earth's atmosphere created a large chemical energy potential, which ultimately allowed organisms to extract energy from organic matter approximately 18 times more efficiently than without oxygen. Without primitive life, Earth's atmosphere would not have contained enough oxygen to support its current life forms, including humans. The main waste products of this oxygen-based respiratory metabolism are water and carbon dioxide. Through a series of symbiotic events, two basic and interdependent metabolic pathways—oxygenic photosynthesis and aerobic respiration—form the basis of all complex multicellular life on Earth. Their evolutionary histories, inferred from gene sequences, are part of the profound record of all life forms having evolved from a few common ancestors.

Metabolism is a universal feature of living systems. All organisms must acquire and transform energy into forms that they can use to make new cells and repair old ones. In the process, all organisms exchange gases with their environment. Gas exchange provides a mechanism to analyze metabolic pathways and fluxes on local and global scales. It is a crucial link between organisms and their environment. When organisms take up energy and resources and expel metabolic byproducts, they shape not only their local environment but ultimately the planetary environment (Frausto da Silva and Williams, 1996; Sterner and Elser, 2002). Although the environmental consequences of an individual organism's metabolism can be small and localized, the metabolic effects of large collections of organisms are global. The planet is habitable for large, multicellular, air-breathing animals like humans only because other creatures have made it habitable. The atmosphere also dissipates heat and buffers temperature, which allows for relatively stable forms of life. Because the metabolisms of organisms are linked, to each other and to the atmosphere and climate, this is an area with potential for theoretical unification.

There are conserved metabolic pathways by which organisms capture, transform, and dissipate energy. This chapter considers the evolution of these pathways and the interaction of energy metabolism with pivotal materials such as carbon and nitrogen. An expansive view of metabolism is taken throughout the chapter. It is considered both at the level of cells and organisms and at the level of ecological systems and the entire biosphere. This multiscale approach is essential given that it is the combined effect of individual organisms' metabolisms, which have the potential to affect regional and global environmental conditions.

An important challenge at the intersection of biology, geochemistry, and physics is to understand how the global metabolic network evolved, what the feedbacks were that led to the constrained variations in gas composition of the planetary atmosphere, and the limitations of these processes on organismal, ecological, and geological spatial and time scales. Understanding this vast global metabolic network requires developing a global "systems geobiology," the root of which lies in the origins of life on Earth and which is deeply grounded in the fundamental physiological pathways of life.

"Systems geobiology," in this report, is defined as the integrated study incorporating geochemistry, geophysics, and other environmental sciences with genomics, ecophysiology, and mathematics to understand the processes and feedback mechanisms influencing Earth's overall metabolism. The further goal is to improve our ability to predict responses of the Earth's systems to external and internal perturbations. This discipline is as new as it is urgently needed and will require significant theoretical investment to tie together its diverse components.

The collective metabolism of human societies, which includes a variety of industrial processes in addition to human biological functions, generates enormous amounts of byproducts. Human activities have altered the composition of gases in Earth's atmosphere, the distribution of water in terrestrial ecosystems, and nutrient regimes in rivers and along continental margins worldwide. These byproducts have consequences for the natural environments that sustain humans and also influence biospheric metabolic processes by modifying the physical environment. Knowledge of the intricacies of metabolism is critical to (1) understand the consequences of our metabolism at all scales, (2) devise and facilitate remediation strategies for the ecosystems that are degraded but crucial to sustaining humans, and (3) reduce the generation of noxious metabolic byproducts and accelerate their safe disposal. The issues addressed by systems geobiology are fundamental in public discourse; these issues include understanding the importance of metabolism for all life on Earth and the extent to which specific metabolic processes can be altered to ameliorate human-caused effects on biogeochemical cycles.

Systems geobiology cuts across traditional disciplinary boundaries. For example, global metabolic fluxes are the cumulative result of the specific capabilities of individual molecules, powering individual cells in different organisms, which themselves interact in many different communities. Research in this area requires expertise in microbiology, enzymology, protein chemistry, cell biology, biophysics, comparative physiology, geochemistry, and ecology, among others. These topics therefore require the combined skills of physical scientists and biologists of all kinds. Practitioners of this new science have to work at many scales that span from genomics to the atmospheric sciences. The broadly integrative training approach of the physiological sciences will likely be invaluable in training students of the Earth's "physiology." Such an interdisciplinary approach is rare in biological curricula. One possible course curriculum would begin with an overview of the metabolic processes of archaea and bacterial cells, outline the evolution of these processes during Earth's history, and conclude with an overview of the biosphere's biogeochemical cycles.

THE ROOTS OF METABOLISM

The ability to acquire energy and convert it to biologically usable forms (energy transduction) depends on a few, virtually immutable, complex molecular machines. These machines catalyze reactions in which electrons are transferred from reduced, high-energy molecules to a small set of molecules that act as energy transfer receptacles. All known energy transduction machines evolved in microorganisms over 2.5 billion years ago and were spread across all domains of life by lateral gene transfer and endosymbiotic events. The relatively free exchange of metabolic machines early in the history of life has resulted in a set of core metabolic pathways shared by all organisms. Although biologists do not have a detailed understanding of how these energy transfer machines evolved on a subcellular level, these shared molecular entities now form an interdependent planetary "electron market" where reductants and oxidants are exchanged across the globe. The scale of this electron market is planetary because gases, produced by all organisms, can be transported around Earth's surface by the ocean and the atmosphere.

From a metabolic perspective, living systems use a relatively small suite of conserved ancient pathways. The vast diversity of metabolic pathways can be divided into sets of metabolic circuits that perform three different functions (Figure 4-1). The first set of circuits is devoted to acquiring environmental energy. Living systems harvest energy from sunlight and from inorganic and organic compounds, and they transfer this energy to electron or hydrogen carriers. The second set of circuits uses the energy of oxidation or reduction (redox) reactions to pump ions across membranes to establish ionic charge gradients. Once an electron is accepted, it flows downhill (in an energy sense) until the circuit is closed by the reduction of an electron acceptor. The charged membrane is a biological capacitor that serves several different functions, among the most significant of which is the synthesis of adenosine triphosphate (ATP), the universal high-energy compound of living systems. A third circuit of reactions serves to employ a source of carbon and energy-rich molecules to synthesize new organic compounds, and thus make new cells and repair old ones. All the anabolic reactions required to manufacture cells and the tissues of multicellular organisms are in this category.

The previous chapter discussed the diversity of life forms and functions that evolution has generated. By contrast, the universality of the genes, proteins, and compounds that participate in these three sets of metabolic pathways is noteworthy (Benner et al., 2002). The molecules that transport energy used by various living systems are, for the most part, the same and seem to be a near-universal feature of life on this planet. Organisms ranging from anaerobic archaea and bacteria to strictly aerobic animals have adopted nicotine adenine dinucleotide (NAD), flavins, and quinones as their energy carriers. Genomic data show that the pathways used to synthesize these compounds can be found across the boundaries of life's domains. The reactions for energy transduction are found in archaea and bacteria and are present in eukaryotes as well, as a result of an ancient endosymbiosis (Nealson and Rye, 2005). The enzymes responsible for the hydrolysis of ATP and the genes that code for them are abundant in most organisms. Those

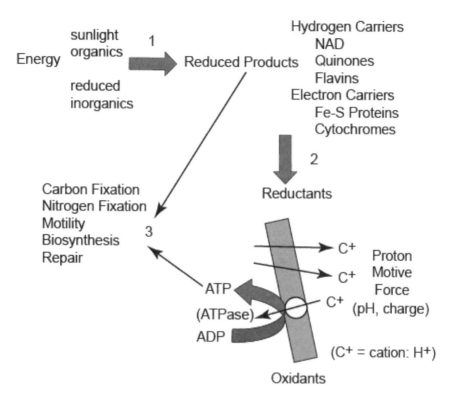

FIGURE 4-1 The three fundamental energy metabolism processes: (1) The formation of reduced products from sunlight, organic molecules, or inorganic reduced molecules. Energy is transferred to reduced hydrogen or electron carriers that are then used directly for anabolic reactions (3), for C or N fixation, or for (2) the generation of adenosine triphosphate (ATP). ATP is generated by the pumping of cations (usually protons, H^+) across a semipermeable membrane to establish a gradient. Many anabolic functions (3) require ATP and/or reducing equivalents.
SOURCE: Reprinted from Treatise on Geochemistry: Volume 8, Biogeochemistry, K. H. Nealson and R. Rye, Evolution of Metabolism, Pages 41-61, Copyright 2005, with permission from Elsevier.

enzymes and genes have been co-opted to perform all sorts of other functions, many of which now have nothing to do with ATP (Saier, 2000).

All three domains of life appear to use similar approaches for energy capture and transduction and the same (or very similar) molecules to fulfill these two functions. Either the last common ancestor of the three domains had already evolved these processes or these processes were widely exchanged by lateral transfer, and the now common processes proved to be better than their alternatives. In either case, it is likely that the evolution of life's redox chemistry (and the diversity of pathways that it now has) evolved very early in the history of life, long before the deposition of the first macrofossils.

There is a global integration of the planet's metabolism. The composition of the atmosphere, and hence the conditions for life, is the result of the balance of inorganic processes and of complementary metabolic processes. The biogeochemical tension between nitrification and denitrification and photosynthesis are good examples. The metabolism of living systems and the physical state of the planet are linked by complex and still poorly understood feedbacks. An important challenge for the future of the biological sciences is to forge collaborations with the geosciences with the goal of understanding the full metabolic network of the planet.

METABOLISM:
A CELLULAR PROCESS WITH GLOBAL CONSEQUENCES

The study of metabolic processes, in all their guises, is a unifying theme in biology. Studying metabolism, the flow of energy and molecules in the cell, at almost any level of organization is a challenging enterprise that demands the development of imaginative conceptual approaches and new technologies (Box 4-1).

One of the promises of systems biology (including the integration of transcriptomics, proteonomics, and metabolomics) is to develop adequate working models of metabolic cell function. Establishing the link between macromolecular structure and metabolic function is also a goal of many disciplines in biology, ranging from organismal physiology to ecosystem and planetary ecology. Indeed, the interdisciplinary approach that organismal biologists use to investigate the function of whole organisms might be the best model for how to teach, study, and communicate metabolism to the public (Feder, 2005).

At the higher levels of organization, the conceptual and technological challenges are as difficult as they are urgent. Humans are exerting a major impact on the metabolic fluxes of the planet. An understanding of biosphere-atmosphere interactions requires integrated input from biology, atmospheric science, and geology. The role and importance of the physiological processes of plants, microorganisms, and animals in both terrestrial and marine environments on the composition and behavior of the atmosphere still need to be fully characterized. The processes that shape biosphere-atmosphere interactions occur at many spatial scales and can take place over decades and centuries. New actors in this system continue to be discovered, like the archaea that are major consumers of methane in oxygen-free sediments (Raghoebarsing et al., 2006).

BOX 4-1
Stable Isotopes Reveal the Global Influence of Metabolism

The observation of enzyme-driven molecular effects at the planetary level reveals the immensity of the magnitude of life's metabolism. Remarkably, these global-scale signals can be re-created in test tubes and in greenhouses. Just as researchers interested in systems biology and in tracking the evolution of biological systems rely on nucleic acids and the polymerase chain reaction (PCR), ecologists interested in measuring the fluxes of energy and materials among components of ecological systems (ranging from cells to the whole biosphere) increasingly rely on stable isotope analyses. Interpreting the stable isotope signals of life's metabolism requires putting together information derived from the study of metabolic processes at levels that range from cells to broad geographical regions.

All the macromolecules that comprise life are composed of six major elements: H, C, N, O, S, and P. Of those the first five have stable isotopes that can be distinguished by their mass. The enzymes that mediate metabolic reactions are often sensitive to differences in the dissociation energies of molecules with different isotopes in them. Consequently, molecules that contain isotopes of different masses are incorporated differentially into the products of incomplete metabolic reactions. For example, in oxygenic autotrophs (organisms that use the energy from sunlight to produce sugars, releasing oxygen as a byproduct), Rubisco (ribulose 1,5-biphosphate carboxylase-oxygenase) fixes CO_2 to make sugars. This enzyme discriminates against CO_2 containing ^{13}C and produces sugars that are greatly depleted in this isotope. Because Rubisco is the most abundant enzyme on Earth and processes enormous amounts of CO_2, there is a substantial accumulation of ^{13}C in organic matter buried in the lithosphere and in the isotopic composition of the atmosphere. On land the fixation of CO_2 varies seasonally and seasonal changes in ^{13}C concentration can be observed in atmospheric air. CO_2 dissolves in the ocean with little discrimination. Therefore, ^{13}C is useful to distinguish CO_2 uptake by terrestrial vegetation and by dissolution in the oceans.

Similarly, during respiration, ^{16}O (i.e., O_2 containing two atoms of ^{16}O) is preferentially used to oxidize organic matter, leaving the major heavier isotope ^{18}O in the atmosphere. In contrast, there is virtually no fractionation of oxygen isotopes in photosynthesis. Hence, variations in the $^{18}O/^{16}O$ ratio provide a geochemical framework to assess how closely coupled photosynthesis and respiration are on geological time scales (Sowers and Bender, 1995). The geological records of S and N isotopes reflect the oxidation state of the oceans and atmosphere, as well as periods when the Earth's systems were greatly perturbed (e.g., through mass extinction events). Understanding how the Earth systems responded to these perturbations and the time scale of recovery is critically important, as human perturbations potentially can force the planetary atmosphere/climate into a new mode, very different from that which humans have experienced since the evolution of *Homo sapiens* ~ 200,000 years ago. This understanding requires an integration of knowledge from the physical bases of isotopic fractionation to the molecular processes responsible for the observed isotopic variations. Biologists will need to be trained so that they are capable of adopting new tools like this that cut across levels of organization and that allow scaling up the consequences of the molecular details of metabolic enzyme function to their global effects.

The study of biosphere-atmosphere interactions, therefore, emphasizes the use of model simulations (Moorcroft, 2006). Nonetheless, these models need to be informed by laboratory experiments that probe the responses of organisms to changes in the atmosphere and by reliable measurements of the relevant gas fluxes. These require sophisticated gas exchange methodologies and remote sensing techniques to scale up what are now, by necessity, sporadic measurements with respect to relevant geographical scales. So far, coupled biosphere-atmosphere models have largely avoided accounting for the diversity of function in the organisms that make up real ecosystems. It is clear that biosphere-atmosphere models cannot yet account for all the details and all the biological structure in ecosystems. However, it might be possible to group organisms into functional groups based on the effect of diversity in physiology with respect to the atmosphere. Indeed, ecologists have now begun to measure variation in the function of microorganisms and plants at the global scales that are appropriate to account for biosphere-atmosphere interactions. Life's metabolism is changing and continues to change and shape the atmosphere. But the atmosphere has also shaped life. To go beyond this simple observation, it is necessary to create a predictive science of the biosphere's metabolism and its effect on the atmosphere. This now stands as a major challenge for biology, helping to advance the science of the Earth's metabolism.

The metabolism inside a cell has profound consequences for the environment in which that cell exists. For example, large-scale heat production is critical for complex behavior in metazoans and is the base for endothermy in mammals and birds. Large-scale heat production also occurs in microbial communities, in termite and ant colonies, and at ecosystem levels, when forests and phytoplankton dissipate large fractions of absorbed solar energy, thereby altering the thermal structure of the local environment (Lewis et al., 1990; Gates, 2003). One mechanism involved in generating heat involves the two primary products of energy transformation: ATP and nicotinamide adenine dinucleotide phosphate (NAD [P]). The former is required for catalysis, macromolecular synthesis, and protein conformation. The latter is required for redox reactions. How ATP/NAD(P)H ratios are controlled at the cellular level is still poorly understood, but the ratio of these two molecules is critical in determining energy transformation efficiency. Excess production of ATP can be coupled to exergonic reactions, thereby dissipating energy as heat.

Photosynthesis appears to have evolved in the early Archean (>3 billion years ago), although exactly when remains unclear. Initially, the process was almost certainly anaerobic; the energy of the sun was used to extract electrons and/or protons from relatively low energy molecules and elements including hydrogen sulfide (H_2S), ferrous iron (FeII), and even preformed organic matter (CH_2O) to chemically reduce CO_2 to form organic matter. By the late Archean and early Proterozoic (approximately 2.5 billion years ago), geochemical data suggest that water (H_2O; a reduced form of oxygen) was oxidized by photosynthetic organisms to produce molecular oxygen (O_2). The burial of the photosynthetically produced organic matter in ocean sediments allowed oxygen to accumulate in Earth's atmosphere. Indeed, without the burial of organic matter—a geologically controlled process—Earth would have remained anaerobic. The slow rise of oxygen through the mid to late Proterozoic altered forever the metabolic networks that subsequently evolved in the first half of Earth's history. Oxygen is an extremely strong oxidant; when coupled to the oxidation of organic matter, it yields up to 18 times more energy than anaerobic metabolism. The evolution of oxygen in Earth's atmosphere and oceans "supercharged" biological metabolism, ultimately facilitating much faster metabolic fluxes of elements through biological systems. The use of oxygen as an electron sink forms the basis of another metabolic pathway—aerobic respiration. That pathway originally evolved in microorganisms and then was captured, through endosymbiosis, by other microorganisms, forming eukaryotic cells. Eukaryotic cells, the basis of all "complex" multicellular animal life, therefore, are the result of the shift of Earth's metabolism to reliance on oxygen.

The vast majority of carbon on Earth is stored in the lithosphere (the outer solid part of the Earth) in approximately a 4:1 ratio of inorganic carbon (carbonates) and organic matter. The organic matter represents a fraction of reducing equivalents (such as electrons) that have been removed through biological metabolism, thereby allowing oxidation of the Earth's surface. On geological time scales the effect of biological metabolism on atmospheric CO_2 may be outweighed by other sources. Significant variation in the CO_2 content in Earth's atmosphere is apparently primarily in response to tectonic activity (Berner, 2004). Nonbiological sources like volcanism can also have a major impact on atmospheric CO_2 content. On a shorter time scale, however, CO_2 concentration in the atmosphere is controlled primarily by exchange with the ocean and the biosphere (Falkowski et al., 2000).

How metabolic pathways adjust to changes in CO_2 on geological time scales remains unclear. Some microbial organisms are able to use several different metabolic pathways. Joshi and Tabita (1996) discovered a common regulatory circuit that regulates the balance between photosynthesis, respiration, and nitrogen fixation within these bacteria. In photosynthetic cells, the ultimate choice of metabolic pathway might lie in how the balance between chloroplasts and mitochondria is controlled (Nisbet and Fowler, 2005). This intriguing conjecture remains to be explored and tested but is of considerable importance to develop an understanding of the complex coupling of atmospheric carbon levels, individual organisms' metabolisms, and the net effect of the metabolisms of entire communities.

Another area ripe for theoretical and conceptual breakthroughs is the role of microorganisms in the metabolism of plants and animals (Box 4-2).

CROSS-CUTTING QUESTIONS IN METABOLISM

Whole-organism metabolism can be described by simple molecular patterns of products, but the processes that shape these patterns remain poorly understood. One of the challenges of biology is establishing clear mechanistic links between structure and function. This challenge cuts across levels of organization. For example, understanding the effects of metabolism on how ecosystems work demands establishing connections between structure and function. Structure at this scale is defined as the composition and abundance of species, and function is defined as the integrated metabolism of a biological community, including respiration, primary productivity, decomposition rates, nitrification, denitrification, and other functions. Biologists have known for a long time that two variables have profound influence on an organism's metabolic fluxes: body size and temperature. The effect of these two factors on a handful of metabolic functions (aerobic respiration) and in a handful of taxa (animals) has been carefully studied. The rate of aerobic respiration is known to be proportional to body size raised to an exponent. Respiration, like all metabolic processes, is also known to be dependent on temperature. The joint effect of body size and temperature on the rate of aerobic respiration can be described by the product of a power function of body size (called an allometric function) and the Arrhenius-Boltzmann equation, which relates the rate of biochemical reactions with temperature (Gillooly et al., 2001).

A recent flurry of theoretical explorations attempts to explain not only the seeming universal dependence of aerobic respiration on body mass and temperature but also the putative ubiquity of the value of ¾ in the exponent of the power function that relates metabolic rate with body size. The theories have led to the conjecture that the value of this power is the consequence of the structure of the systems that distribute oxygen and nutrients in organisms (West et al., 1997). The theory has been extended to terrestrial vascular plants and has led to the remarkable prediction that both photosynthetic rate and respiration should also scale with plant mass to the ¾ power. This theoretical research has been accompanied by attempts to include these relationships in scaling exercises that predict ecosystem-level properties such as the metabolic balance of the oceans and the productivity and decomposition rates in terrestrial ecosystems (López-Urrutia et al., 2006). These calculations suggest that first-order estimates about the magnitude of these processes can be made from knowledge about the size

BOX 4-2
The Role of Microbial Communities in Metabolism

To elucidate the structure of a biological community, biologists need to determine how many species there are and their relative abundances. Answering these questions is more difficult than it seems, especially for the microscopic organisms that constitute the metabolic backbone of most ecosystems. A large fraction of these microorganisms cannot be characterized by the traditional approach of isolation and culture. The techniques of metagenomics, as discussed in Box 3-2, make it possible to probe the structure of microbial communities and to link that structure with metabolic function.

Metagenomics is also making it possible to explore the microbial communities that live on and in virtually all higher organisms. In fact, our theoretical understanding of metabolism now must incorporate the realization that all organisms—from plants to invertebrates to mammals—have an associated microbial community that affects many aspects of their physiology, including metabolism. The analysis of the metagenomes of various host-associated microbial communities has been used to diagnose functional metabolic differences among communities (Tringe et al., 2005). It also can be used to assess the spatial heterogeneity in metabolic function in a single community and its changes through time. For example, differences in the nutritional physiology among individuals of a single species might be shaped by differences in the metabolic capacities of their nutritional symbionts. Ruth Ley and her collaborators (Ley et al., 2006) recently found large differences in the composition of the microbial community of obese and lean people. In an accompanying study by the same group, Paul Turnbaugh (Turnbaugh et al., 2006) transferred the microbiota of obese mice to lean, microbe-free mice. These recipient mice extracted more calories from their food and gained slightly more fat than mice receiving microbiota from lean mice. Although these results should be interpreted cautiously (cause and effect are unclear), their results suggest that differences in the efficiency of caloric extraction from food might be determined by the composition of the gut's microbiota. More generally, the results emphasize how the metabolic capacities of multicellular animals and plants are complemented, and sometimes extended greatly, by those of their symbiotic microbes. The implications of this complementarity are profound not only for the study of metabolism but also for the study of health, development, and evolution. There are many opportunities for the development of theories to explain and predict the impact of these microbial communities on multicellular organisms. Metagenomic surveys are one of the novel genomics-based approaches that will allow the empirical testing of such theories, opening up intriguing lines of scientific inquiry that will involve microbiologists with nutritionists, physiologists, and ecologists.

Current technologies have been sufficient to show the tremendous promise of metagenomics approaches, but detailed understanding of microbial communities that may contain thousands of species will require significant advances. Most importantly, theoretical advances in understanding what controls community assembly, structure, and stability will be very important in guiding what technology to develop and what experiments to do.

distribution of the organisms that structure these ecosystems and from the temperature at which they operate.

What has been called the metabolic theory of ecology has many detractors. The details of its theoretical foundation have been criticized, and the ubiquity and universality of the ¾-power rule have been challenged. Furthermore, the patterns revealed by the approach can have large errors around predicted values. Although the theory undoubtedly has limitations, it is a bold and promising attempt. The ubiquity of the power functions relating metabolic function to body size and the importance of temperature for biological processes are undeniable. The challenge is not to dismiss the theory but to test it, find the cases in which it fails, and modify and strengthen it to improve its power. The task is to explain not only the general trends in these size-temperature-metabolism relationships but also the details that make some systems deviate from them. The metabolic theory has to be linked to the molecular details of the metabolic architectures of living systems. The magnitude of an organism's metabolism is the outcome of feedback between the "the whole," construed as the whole organism, and "the parts," construed as the cellular and subcellular machinery. Does a mitochondrion "know" that it is within a mouse or an elephant and behave differently in a predictable manner? Since a large fraction of the functional proteins come from the nuclear genome, there is a strong basis for such coupling. The nature of these feedbacks remains unclear and is a fertile area of investigation. A complete metabolic theory would link the details of the structure of metabolic pathways in cells and organelles with the "macro" patterns that biologists can discern and that the metabolic theory aims to explain. Biologists are still far from this goal. Although a metabolic theory of life is still being constructed, the metabolic theory of ecology has forced biologists to attempt to search for simplicity in the patterns produced by seemingly complex processes and has given a glimpse of hope about the feasibility of the task.

Although the conservation of metabolic pathways is clear, some surprises have emerged in recent years in relation to size-temperature-metabolism relationships. One is the discovery of extremely slowly growing microorganisms deep in Earth's sediments (D'Hondt et al., 2002). Those organisms grow so slowly that it is virtually impossible to measure gas exchange with their environment, yet they are not small compared to other microorganisms. In contrast, the discovery of deep-sea hydrothermal vents revealed the presence of symbiotic chemoautotrophic bacteria associated with numerous invertebrates (animals), living near very high temperature, sulfide-rich waters emanating from marine volcanoes. Those two examples suggest that the temperature-size-metabolic rate relationships derived from observations of metazoans and higher plants probably cannot be extrapolated to microbial communities. Yet the basic thermodynamic processes that control metabolic rates viz. temperature still apply. Clearly, the "anomalies" suggest that the patterns so far discerned are not universal and that investment in further theoretical work to understand these relationships would be highly productive.

CONCLUSION

Systems geobiology and such approaches as metagenomics cut across traditional disciplinary boundaries. Answering the questions they pose requires the combined skills of physical scientists (including computer scientists, physicists, and chemists) and biologists. For example, global metabolic fluxes are the cumulative result of the specific capabilities of individual molecules, powering individual cells in different organisms, which themselves interact in many different communities. Thus, a deep understanding of these fluxes will require input from fields as diverse as enzymology, protein chemistry, cell biology, biophysics, comparative physiology, and ecology, just to list some of the necessary biologists' areas of expertise. Understanding the biosphere's metabolism will require that technologies developed to measure metabolic processes at small scales be refined and scaled up to appropriately broad temporal and spatial scales. These technologies include combining stable isotope and gas-exchange measurements with remote sensing and mathematical modeling. So far, the study of Earth's metabolism has been based on the theoretical frameworks provided by thermodynamics and the laws of chemical equilibrium. The next chapter discusses the problem that in order to understand how cells really work, biologists will need to complement these frameworks with other approaches that are more realistic at the scale of cells. It may be that a more accurate understanding of cell metabolism will contribute to a better understanding of the planet's metabolism. This observation emphasizes that the practitioners of this new science have to work comfortably across scales—from genomics to the atmospheric sciences through cell and organismal biology. The broadly integrative training approach of the physiological sciences might be invaluable to train students of Earth's "physiology," such as the biological, geological, and atmospheric processes that facilitate global biogeochemical fluxes of elements and maintain this planet far from thermodynamic equilibrium.

Deep understanding of the processes and interactions that couple the biosphere and the geosphere has tremendous potential to generate solutions to societal problems. For example, one question that could be addressed is "Which key biological reactions, if catalyzed on an industrial scale, would make the transition to sustainability?" Clearly the photochemical splitting of water would potentially provide hydrogen as an infinite energy carrier, and therefore can potentially negate the need for combusting fossil fuels.

Similarly, the ability to fix N_2 efficiently would alter the impending crisis of eutrophication of coastal waters throughout the world.

Over the past century, humans have dramatically altered the global environment, extracting resources and energy to facilitate economic growth and development. Many valuable resources, such as fixed inorganic nitrogen and organic carbon, are essential for production of food and for fuels. These biologically critical molecules are either produced inefficiently by chemical synthesis or are not available in sufficient quantities.

Over the next century, a major challenge for society will be to develop or redesign metabolic pathways, based primarily on microbial systems, to greatly accelerate fluxes of materials and energy. One of the major outcomes of understanding metabolic pathways and energy transformation processes is to replace technologies designed in the 19th and 20th centuries with sustainable processes that are biologically driven.

5

How Do Cells Really Work?

Inside each microscopic living cell, thousands of diverse chemical reactions must take place at the right time, in the right places, and in the right order. Scientists can re-create many of these individual reactions, or even a few coupled reactions in the laboratory, but the spacious and uniform conditions of a test tube bear little resemblance to the crowded and highly structured interior of the cell. The sequestration of chemical reactions within cells was probably one of the critical factors in the early evolution of life. Understanding how this complex milieu developed and varies across different life forms will serve as another profound illustration of the ways that evolution has maintained certain common ancestral features throughout life's history.

The theoretical frameworks provided by thermodynamics and the laws of chemical equilibrium have been used productively to study the chemical reactions of life. However, analysis of individual molecules within cells, tissues, and developing embryos reveals important differences from studying these molecules in aqueous solution. To understand the behavior of even familiar macromolecules, biologists need to study them under the conditions found within cells and tissues. These conditions differ from those of typical biochemical experiments in that reactions do not proceed to chemical equilibrium, reaction volumes are small, solutions are crowded and inhomogeneous, the concentrations of enzymes are often higher than that of their substrates, and many reactants are immobilized on membrane or proteinaceous surfaces.

UNDERSTANDING THE CELL

Cells and organisms are crowded and complicated (Figure 5-1). Although the macromolecular structures within cells must be self-assembling, they perform this self-assembly following elaborate temporal expression and spatial localization of their individual constituents. The highly energy-requiring events of a living cell require the synthesis of large macromolecules and their specific localization against concentration gradients and through crowded solutions. The flux of energy that moves through the system allows the reactions to be held away from equilibrium, and this is an essential characteristic of living systems. Understanding these processes requires an appreciation of nonequilibrium thermodynamics, a situation that can be sustained when energy is constantly added, as it is during life.

In chemistry one of the most powerful, unifying concepts is equilibrium thermodynamics, the principle that allows a prediction with confidence that ice left at room temperature will melt and that water put in a freezer will become ice. Under any given set of conditions, a chemical system will tend to change its properties, including temperature, pressure, and concentration of reactive chemical species, toward a particular stable state called "equilibrium." If the system is perturbed slightly away from its equilibrium state, it will robustly return to equilibrium. If the system is left alone, it will remain at equilibrium indefinitely. There are excellent, accurate mathematical formalisms for calculating and predicting equilibrium states of even

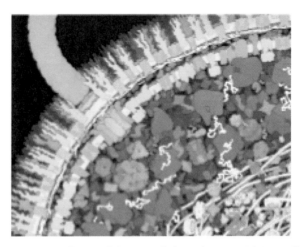

FIGURE 5-1 Artist's rendition of the crowded conditions within a cell. Illustration by David S. Goodsell, Scripps Research Institute.

very complex physical and chemical systems, using a fairly small number of "state variables."

Most of biologists' understanding of the biochemical reactions in living cells has come from experiments with purified enzymes and substrates studied in isolation. For the most part, these experiments have been done under conditions that are not mimicked within living cells and interpreted using assumptions that are not appropriate for living cells. For example, an isolated, purified enzyme placed in a test tube with its substrate will catalyze the conversion of the substrate into product until equilibrium is reached. However, the actual behavior of the enzyme and its substrate inside a living cell is, in most cases, different for several reasons, including, for example, that the products of reactions are usually then further transformed. Living cells do not operate at chemical equilibrium. A cell or organism at equilibrium would be dead. Additional assumptions that are appropriate for reactions in dilute aqueous solutions but not in cells and tissues are:

- that reaction volumes are infinite;
- that the solutions in which reactions occur are dilute, well defined, and homogenous;
- that molecules collide due to diffusional motion; and
- that concentrations of substrates are higher than concentrations of enzymes.

The inappropriateness of these simplifying assumptions for all living systems is discussed in this chapter and underlines the need for new approaches to better understand the biochemistry of the living cell.

A DIFFERENT VIEW OF CELL CHEMISTRY

In cells, reaction volumes are finite. Cell volumes vary—for example, a mycobacterium is about 10^{-11} microliters and a *Xenopus laevis* egg is about 0.5 microliters. In the bacterium *Escherichia coli*, a protein present at 1 nanomolar concentration, a concentration at which many enzymes are studied in the laboratory, is calculated to be present at 0.6 copies per cell. The volume within a 50-nanometer vesicle such as the vesicles involved in protein transport within cells is 6×10^{-20} liters, which means that one free proton within such a small vesicle would yield a pH of 6.0, and 50 free protons would yield a pH of 5.0. Therefore, only a few events would be required to acidify a vesicle involved in endocytosis, for example. Furthermore, cells vary greatly in shape, from roughly spherical to elongated. Cells are highly organized and many are specialized for specific functions. In some cases structure obviously correlates with function, such as in the polarization of epithelial cells.

In cells, solutions are not dilute, well defined, or homogenous. Instead, the interior of a cell is 17 to 35 percent protein by weight, and the individual proteins are likely to be associated with each other and with other molecules in complicated ways. The effects of this macromolecular crowding are extensive. Mobility of molecules, including water, is decreased, with larger molecules more affected than smaller molecules. "Nonspecific" interactions between proteins—such as those interactions that do not normally occur in aqueous solutions—are enhanced, and "specific" interactions occur more readily. For example, since chemical equilibrium theory is based on activities, not concentrations, the equilibrium constant for dimerization of a 40,000-kilodalton molecular weight protein is 10 to 40 times greater in a cell than in dilute solution, and its tetramerization is 1,000- to 100,000-fold greater. The ability of cellular machinery to localize individual proteins and other macromolecules within the cell in specific ways will lend apparent specificity to otherwise nonspecific reactions. For example, there are many different pairs of "SNARE" proteins (Figure 5-2) that are known to facilitate fusion between intracellular membrane compartments. Within cells the many different membrane fusion events that are facilitated are highly specific, with endosomal membranes fusing with lysosomal membranes and not, for example, with mitochondrial or

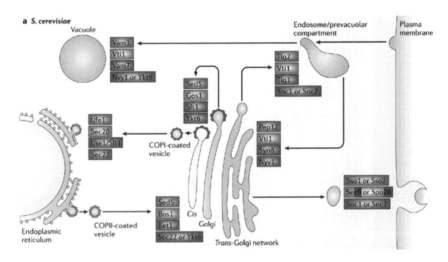

FIGURE 5-2 Much of the specificity of SNARE-SNARE fusion is likely to derive from specific cellular localization. Reprinted by permission from Macmillan Publishers Ltd: Nature Reviews Molecular Cell Biology 7:631-643. SNAREs—Engines for Membrane Fusion, R. Jahn and R. H. Scheller, copyright 2006.

nuclear membranes. However, purified SNARE proteins mediate fusion reactions relatively promiscuously, showing affinities for each other that do not correlate with their known partnerships within the cell. A conceptual framework based on results in aqueous solution would lead to a search for additional specificity "factors"; in this case, however, it appears that the specificity most likely resides in the intricate mechanisms of the localization and orientation of these disparate SNARE proteins within the cell. Therefore, development of conceptual frameworks that take into account the crowded interior of the cell and guide experimentation to determine how organization and localization are achieved has great potential.

Within cells, diffusional motion is highly restricted. Movement of groups of individual molecules can be measured inside cells by photobleaching fluorescent molecules in a limited area and monitoring the rate at which they exchange with unbleached molecules. These and other measurements have revealed that proteins move 10 to 50 times more slowly inside the cell than in aqueous solution, with many proteins displaying a completely immobile subpopulation. These restrictions to movement are strongly size dependent, with larger complexes being almost immobile. What limits the diffusion of these molecules? Is it nonspecific interactions with the high concentration of other proteins, or specific interactions that cause many proteins to function in much higher-order complexes than previously suspected, or sieving through the network of the cytoskeleton, or reduced water activity due to macromolecular crowding? To the extent known so far, all these factors come into play. Poor diffusion of molecules within cells necessitates that any specific intracellular localization must be accomplished by specific transport of the mRNAs that encode the proteins, the proteins themselves, or both.

In cells, concentrations of enzymes are often higher than their substrates. The concentrations of many steady-state metabolites are lower than the measured binding constants for the enzymes that process them, predicting that there should be little free substrate. How, then, are multistep reactions accomplished? "Substrate channeling" is a common solution (Figure 5-3). From carbamoyl phosphate synthetase to transfer ribonucleic acid (tRNA) synthetases, enzymes that catalyze individual steps of multistep reactions have been found to be co-localized or present in large complexes. Such complexes might, in fact, exclude nonchanneled substrates. In the case of tRNA synthetases, the direct introduction of free tRNA into cells does not result in its incorporation into charged tRNA synthetase, even though such reactions occur readily among purified components in aqueous solution.

The reality of the inside of a living cell, which has poor diffusion, immobilized reacting groups, and high degrees of localization, changes the

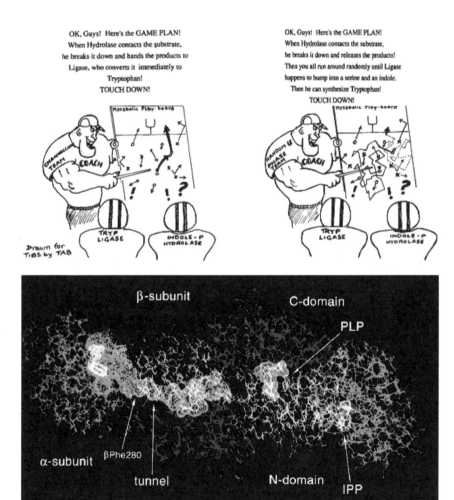

FIGURE 5-3 Substrate channeling.

Upper image: Cartoon depicting substrate channeling in tryptophan synthase. Reprinted from Trends in Biochemical Sciences, Vol. 17, J. Ovadi and P. A. Srere, Channel Your Energies, Page 3, Copyright 1992, with permission from Elsevier.

Lower image: Structure of the tryptophan synthase complex, which the substrate tunnel highlighted. SOURCE: The Molecular Basis of Substrate Channeling in Journal of Biological Chemistry, Vol. 274, by E. W. Miles, S. Rhee, and D. R. Davies. Copyright 1999. Reproduced with permission of American Society for Biochemistry and Molecular Biology via Copyright Clearance Center.

outcomes of interactions between molecules. Take, for example, the well-studied example of an RNA polymerase transcribing a messenger RNA from a double-stranded DNA template. Even in solution, it is unlikely that a polymerase molecule, tracking along the template DNA strand, actually follows a helical path around the DNA molecule. In complex situations—for example, when a newly synthesized strand of RNA becomes associated with ribosomes, or when there are trailing peptide chains, or when dealing with spliceosomes and their complex machinery—it is highly unlikely that these dangling structures would twist around the DNA as the polymerase follows such a helical path. Instead, the DNA template is pulled through a relatively immobile polymerase, removing its helicity as it goes. Therefore, positive supercoiling (an overabundance of helical turns) accumulates behind the transcription complex and negative supercoiling in front of the complex. Conceptually, these turns could be easily removed, especially in a linear DNA template, by diffusional forces that allowed the DNA to spin on its long axis, much as one can unwind an overtwisted telephone cord by allowing the handset to dangle. Nevertheless, within cells, even for linear DNA molecules such as the 40,000 base pair T7 DNA phage genome, transient positive and negative supercoiling occur concomitantly with transcription. In living cells, enzymes termed "topoisomerases" are required to solve these problems during transcription. What prevents the diffusional release of DNA underwinding and overwinding within a cell? Two possibilities are the association of the nominally free DNA ends with subcellular structures or macromolecules that bind along DNA in a manner that is independent of the DNA sequence. Are restrictions to diffusion within the cell so severe that even the spinning of DNA molecules along their long axes is limited?

In addition to the complexity of the environment and reaction pathways of the molecules in a cell, we know that individual reactions are embedded in networks of reactions. The "metabolic network" of a cell is a term now used to describe all of these activities and interactions. It is remarkable, in the light of the foregoing discussion of the complexities of the chemistry of the cell, that the reactions can be intermeshed so beautifully, using substrate channeling as well as other yet unknown mechanisms.

Enzymes and other proteins are often localized within cells, either via specific association with a membrane-bound organelle such as the mitochondrion, endoplasmic reticulum, or the membranous vesicles shown by freeze-etch electron microscopy in Figure 5-4 or via other less understood processes such as association with proteinaceous assemblages such as "P bodies" or "nuclear speckles." For human hepatocytes, the intracellular area presented by internal membranes is approximately 50 times the area of the plasma membrane. The cytoskeleton, even when undecorated with auxiliary proteins, is expected to present another set of surfaces greater than that of the plasma membrane. What are the consequences of the

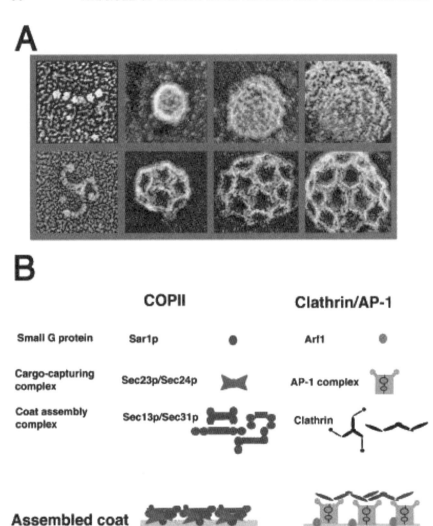

FIGURE 5-4 (A) Surface structure of a COPII vesicle, involved in secretory traffic from the endoplasmic reticulum to the Golgi apparatus, compared to the clathrin coat of a vesicle involved in endocytosis. Deep-etch platinum shadowed electron microscopic images are shown. (B) Images can be used to reconstruct the iterative molecular structures that form their surface coats.

Permission granted by Randy Schekman. Surface Structure of the COPII-coated Vesicle. Proceedings of the National Academy of Sciences, USA 98:13705-13709. K. Matsuoka, R. Schekman, L. Orci, and J. E. Heuser. 2001.

localization of cellular substituents on the surface of such structures for enzymatic activity and specificity? Are the assemblages found within cells merely storage forms of the enzymes of interest, as is often speculated? Or is it possible that ordered arrays of reagents, in the nuclear matrix, on the surface of membranous vesicles or on the exposed surfaces of proteinaceous arrays provide the advantages of surface catalysis to biological systems?

In mature and developing organisms, local interactions among cells are mediated by complex local context. For example, in a developing *Drosophila* embryo, extremely high local concentrations of morphogens are formed and shape cell differentiation and mobility. Developing neurons will establish synaptic connections in response to subtle gradients. Understanding these cues requires not only identifying all the molecules involved but also developing analytical interpretive theories for their roles and testing those theories with, for example, high-definition and quantitative visualization techniques.

CONCLUSION

Understanding the activities and specificities of molecules and larger arrays within cells and tissues will require additional techniques from biophysics, microscopy, materials science, microfluidics, and computational biology. A particular need is the development of microscopy that bridges the gap between fluorescent light microscopy and electron microscopy. In addition to technological advances, the use of simulations of the movements of individual molecules (for example, using the Monte Carlo approach) and the development of theories that incorporate the nonequilibrium conditions of the cell could fuel new scientific advances.

6

What Are the Engineering Principles of Life?

In order for a space shuttle to launch into orbit, dock with the space station, and return safely to Earth, thousands of highly trained individuals and countless sophisticated machines, computer programs, and communications devices need to be engineered, tested, and coordinated. When an orchestra plays a symphony or a basketball team executes a perfect last-second play, each of the participants has dedicated years of training, practice, and discipline. These achievements are examples of human skill, ingenuity, and application of knowledge. Some characteristics of these quintessentially human enterprises are mirrored in basic biology, and nature is full of examples of complex outcomes that result from the coordinated behavior of many simple parts. Across many fields of biology—from the organization of the cell, to the development of multicellular organisms, to the function of the brain, to the group behavior of insects and birds, to the response of ecosystems to environmental change—complex coordinated phenomena are seen to arise out of interaction of a myriad of components. The engineering principles that make possible a space shuttle can be encapsulated in an engineering textbook. Is it possible that there are similarly fundamental principles governing the organization of dynamic interacting systems that hold across all scales of biology? The key to understanding such organizational principles will involve developing a theoretical basis for how biological entities generate aggregates of higher complexity: that is, the constructive principles of biological organizations. Advances in understanding of these biological systems is an especially promising area of research in biology that could have immediate consequences for the understanding of organisms and further applications to complex, human-engineered systems.

An alternative view of engineering is that the field deals with solving constraints or understanding constraints imposed by the characteristics of the parts or organizational structures. Therefore, understanding or developing a theory of constructive engineering principles of life will also yield insights into limits and constraints of biological systems.

The previous chapter discussed how the interior of the cell is highly organized. In fact, much of nature is highly organized, and the organization, or regularity, often seems to emerge without any external direction. A single fertilized egg develops into a mature multicellular organism with all of its many organs, limbs, and blood vessels in the right places. An ecosystem damaged by fire gradually returns to its original mix of species, reorganizing the interdependent community. This chapter will explore the common organizational characteristics and constructive principles of biological systems that lead to complex behaviors, products, and processes.

CORE CONCEPTS

Some core concepts that link different kinds of complex systems are *modules, nodes, networks, emergent behavior, topology (or architecture),* and *robustness.* Table 6-1 provides definitions of these terms and gives examples from several different kinds of systems.

A brief caveat is in order. In this chapter, the terms "modularity," "emergence," and "robustness" will be used to describe characteristics of biological systems that arise at different scales and are in need of further conceptual development. However, the terms have been used in other ways in different domains. However described and however generalizable they may be, the phenomena of modular organization, complex ensemble behavior that might be called emergent behavior, and robustness in biological processes exist and can be described and measured. Whether the best approach will be classical, using existing tools, or whether an entirely new set of formalisms will be required, the problem remains that effective conceptual and theoretical treatment of those topics is not yet available. A satisfactory description or computation of those phenomena is a critical challenge for the future of biology.

Characteristics of Modules

In every biological organization certain divisible parts are recognizable whose repetition and elaboration seem to generate the whole. These parts are often recognized as physically distinct units—the canonical example being the individual organism. In some cases, such units had a conceptual existence before their physical manifestation was known. An example is "the gene" as described before the development of the chromosomal theory

TABLE 6-1 Core Concepts Describing Complex Systems

Term	Definition	Examples
Modules/nodes	Integrated units that can be combined in many ways	Musicians in an orchestra Genes in a developmental pathway Neurons in the brain Locusts in a swarm
Networks	Systems of connected nodes	Orchestra Regulatory feedback loop Brain Swarm Food chain
Topology (or architecture)	The way the nodes are connected in a network	Hierarchical (all individuals connected to one leader) Scale-free (some individuals connected to lots of others, most connected only to a few others) Distributed (individuals connected to neighbors)
Emergent behavior (or properties)	The output of a network	Music Development of a limb or an eye Memory/thought/perception Migration Community
Robustness	Ability of the network to provide the same output despite internal (e.g., the loss of some modules) or external changes	Many different orchestras can play same music; orchestras can play many different pieces of music Many genes can experience mutation but limb or eye still develops normally Some neurons die and most brain activity continues normally Swarm travels despite death of individual locusts or geographical barriers Communities continue despite extinction of some species

of inheritance. Genetic units were defined by certain abstract properties such as segregation of phenotypes upon genetic segregation without knowledge of their physical embodiment. In other cases, loose collections can sometimes be considered a unit with respect to some process or function, as in, for example, a population of individuals that is spatially dispersed can act as a module in an ecosystem. "Module" is a term that seems to capture the sense of these biologically relevant units. While the term is not completely precise, it captures the notion of components or parts organized into larger units that are integrated within and independent (or dissociated) of other units. Furthermore, these modules are generally finite in variety, have properties of superposition such that multiple modules can be combined, and have a certain uniformity to external interface (like bumps on a Lego block) such that a module A that is part of a larger complex can be swapped with module B (Bolker, 2000; Winther, 2001; Schlosser and Wagner, 2004). Such composition may be physical, but the concept of interchangeability within populations (or "demographic replaceability") is an important aspect of modularity that connects modules to evolutionary dynamics (Wagner, 1996). Finally, modularity, like other architectural principles, also constrains or determines the limits of design and function of evolved biological systems.

Examples of modular organization are found at all levels, from substructures of proteins or RNA (Ponting and Russell, 1995; Corbi et al., 2004; Pasquali et al., 2005; Del Sol et al., 2007), to assemblies of proteins that seem to be comprised of surprisingly small numbers of components in a wide variety of combinations (e.g., Devos et al., 2006), to cellular organizations in brains (Redies and Puelles, 2001), to anatomical structures (Raff, 1996; Yang, 2001) and regulatory or metabolic function (Magwene, 2001; Segrè et al., 2005), to classic ideas of ecological communities (Clements, 1936) and even abstract processes such as cognition (Barrett and Kurzban, 2006). Across these scales and substrates, modular organization has been described in terms of physical structure (e.g., anatomical parts or macromolecular geometry), biological function (e.g., cognitive processes), component interactions (e.g., protein complexes), temporal processes (e.g., metabolic flux or development), and genetic architecture. Modularity at these different levels is sometimes coincident, for example, a modular protein domain may carry out a modular function, while at other times no such correspondence can be found. It is an open question the extent to which the concordance of modularity at these different physical and functional levels is promoted by evolutionary dynamics (e.g., Cheverud et al., 2004; Snel and Huynen, 2004).

Characteristics of Interfaces Between Modules

Biological modules typically are made up of other modules at a smaller scale. For example, a cell (itself a module) contains various structural, metabolic, and gene regulatory networks as modules, which in turn contain proteins and RNA molecules as modules. Therefore, the critical aspect of a module is that it might contain many internal parts whose interactions and dynamics are extremely complex, but it has a defined and finite "external" interface that can be connected to other modules. A cell's internal physiology and structure might be dynamic and complex, but to other cells what matters are the cell membrane and interface components such as receptors, transporters, and junctions (Bonnefont et al., 2005). The internal dynamics of a module and the interaction of its parts could be both cooperative and antagonistic and both optimized and random; the parts could also be ephemeral, experiencing constant turnover. In fact the degradation or death of individual modules, while the overall function is maintained, is another universal theme of biological organization. The critical point is that modules should show external coherence, independent of internal complexities. In addition to an invariant external interface, the interfaces should be finite in kind—similar to the finite number of interfaces on a computer such as the USB interface. Sharing a uniform or finite interface (e.g., the phospho-diester bond in nucleotides), especially across functionally distinct modules such as the mitochondrion and the Golgi, allows exchangeability of the modules (Del Sol et al., 2007; Pereira-Leal et al., 2007). The ability to exchange modules creates the possibility of generating combinatorial complexity. For example, during development, gene regulatory feedback loops that have the property of driving cells into a new developmental stage can, through evolution, be linked to other developmental modules to implement major phenotypic changes. In Box 6-1 an example is given of a regulatory loop preserved in star fish and sea urchins but which in sea urchins has evolved to link to another module that drives the development of a skeletal system.

A module as described here is made up of interacting parts, which together interface with the external environment. Variations of the questions "How are such interacting ensembles constructed?" and "How are they maintained?" are found in all subfields of biology. Enumerating the composition and interaction of parts in a cell, in an organ, in a population, and in a community are classic research programs. What varieties of RNA are in a cell and how do they interact with the DNA genome? What are the different types of neurons constituting a hippocampus? How many different species of bacteria make up a gut community? Such inquiries might be considered an essential part of the classic reductive research paradigm, the goal of which is to use the enumeration to build a constructive understanding of emergent properties from the bottom up. Attempts at a constructive understanding of the combined action of the parts lead to the next level functional or interrelational questions: Are all the entities essential? Do the entities segregate into functional groups? What types of interactions are present and, at an abstract level, what is the network topology of their interactions? What are the forces that maintain the ensemble through dynamic changes? Although the research program of characterization and assembly of parts is classically reductive, from a modular perspective, these questions or approaches clearly apply throughout the scales of modular hierarchy—from molecular parts to ecosystems. To put it broadly, a key conceptual challenge is to ask whether there is a common theoretical framework to how modules are created, maintained, and disposed of at all different scales of organization.

Because modules are ensembles of interacting components, it seems intuitive that cooperation is critical to generating emergent properties from the interacting entities. For example, a single individual must be able to perform many tasks to succeed in a changing environment. Sometimes, however, different individuals, and even different species living together, perform different functions in seeming cooperation. Cooperation can include complex social interactions such as division of labor among organisms with complementary metabolic abilities, the provision of shelter, resource gathering, reproduction, and dispersal (Box 6-2).

Even within interacting parts that form a coherent whole (such as a module), competition or antagonistic interaction may also be an essential force. For example, many gene regulation processes within a cell involve antagonistic interaction of two regulatory proteins competing for the same space on the DNA. Some models have suggested that learning and cognition involve competition among neurons and pruning of connections during early development (Rakic et al., 1994). Food webs are an essential part of a community structure and involve antagonistic relationships. Some theoretical models (Livnat and Pippenger, 2006) suggest that internal conflicts might be an essential component of system optimization and that there might be optimally selected levels of modular integration (Hansen, 2003). In some sense, categorization of component interactions into cooperation or antagonism may reflect an anthropomorphic point of view; scientists can choose to describe the interaction of two proteins competing for the same DNA location as antagonistic. From a control system point of view, however, this is simply one way to implement a bi-stable switch. Is a stone arch held up by the cooperation of appropriately molded stones or by the antagonistic opposing forces acting on the keystone?

Similarly, the participation of two species in a mutualistic relationship can be characterized as cooperative or as a tough and ongoing negotiation. For example, in the mutualistic interaction between rhizobia (a group of nitrogen-fixing bacteria) and legumes (including such vegetables as peas), the extent to which the bacteria supply nitrogen to the plant and the plant supplies carbon to the bacteria is just beginning to be approached using cooperative game theory (Akcay and Roughgarden, 2007). In another example, the quintessentially "cooperative" act whereby maternal and paternal genomes are combined through reproduction also includes a competitive element (Haig, 1993).

Thus, one key conceptual question is whether a unified framework for understanding the dynamics of components can be constructed not so much in terms of proximal quality of interactions (such as cooperation

---

Box 6-1
Comparative Network Architecture

The genetic networks regulating the development of the embryonic endo-mesoderm in two echinoderms, sea urchin and starfish, provide an example of comparative network architecture. The regulatory network for endomesoderm development in the sea urchin has been worked out in significant detail (see Hinman below) allowing comparison of the same set of genes in a related but long diverged lineage. The sea urchin *S. purpuratus* and the starfish *Asterina minita* diverged from their common ancestor more than 500 million years ago. The endomesoderms in these animals develop similarly except that the sea urchin has a cell lineage that develops into a prominent skeleton that is entirely lacking in star fish. When a set of key regulatory genes for endomesoderm development was examined in starfish, a three-gene feedback loop that is a key component of the sea urchin system, was found to be almost unchanged in starfish. The conserved circuit has been preserved since the Cambrian era in both lineages.

The structure of the basic developmental circuit is remarkably conserved. The five genes in the regulatory circuit are wired together in essentially the same way. But there are a few key changes in the circuitry. For example, the sea urchin has an autoregulatory loop (of the Krox gene) not present in the starfish, while the GataE gene is auto-activated in the starfish and not in the sea urchin. Also, the FoxA gene represses the GataE gene in starfish but not in sea urchins. These three changes, indicated as red lines in the figure, represent divergences of the two circuits. The major difference between the two systems is a major rewiring of the external connections of this circuit. In the sea urchin the Tbr gene is not connected at all to this circuit. Tbr is used entirely in the skeletogenic network in sea urchin, a function not present at all in starfish. In the starfish the Tbr gene is still there but is regulated by this circuit through connections (shown in red) to three genes—Otx, GataE, and FoxA—that are not present in the sea urchin.

---

or antagonism) but in terms of how such interaction contributes to the control architecture of maintaining the whole module. Understanding how cooperation and competition may be viewed as two sides of the same coin poses a conceptual issue whose resolution offers the prospect of greater understanding of the ecology and evolution of mutualism among species.

In light of modularity, can concepts of population genetics and evolutionary change be modified? What are the levels of modularity at which natural selection can and cannot act? Evolutionary progress depends on some aggregate of modules. This situation requires thinking beyond individual selection—or what has been called the problem of "levels of selection" (Buss, 1987). Evolution in the context of teams or coalitions—that is, ensembles of modules—would apply to the many organisms that forage, evade predators, and reproduce in the context of teams within a social system (Roughgarden

This example illustrates that developmental circuitry can be conserved over very long periods of time, but that it can be modified by evolutionary processes in several ways—connections can be gained and lost. This will alter the computation that is made by this circuit but only in small ways. The circuit can also be rewired to drive an entirely new function by adding connections to the *cis*-regulatory region of genes in other developmental networks (Hinman et al., 2003).

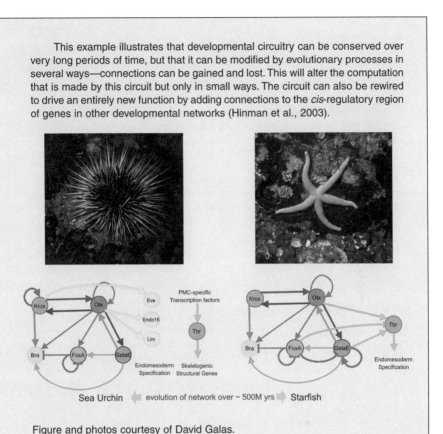

Figure and photos courtesy of David Galas.

et al., 2006). Understanding the relation of team selection to individual selection, together with the adaptive formation and dissolution of such teams, poses a major conceptual challenge for the future.

Because networks of modules are often embedded within networks at higher scales (networks of networks), mathematical tools such as nonlinear dynamics and numerical simulations are critical to understanding how these biological systems depend on the properties of their components. This can be seen directly in neuroscience, where the intrinsic electrical activity of individual neurons depends on the number and kind of voltage-dependent currents, and simulations are crucial for understanding how the properties of the currents alter the excitability of the neurons. Likewise, simulations and mathematical analyses are crucial for understanding how the behavior of networks of neurons is influenced by changes in synaptic strength. At still

---

Box 6-2
Cooperative Behavior of the Slime Mold *Dictyostelium discoideum*

The slime mold *Dictyostelium discoideum* is one of the best-studied examples of cooperative behavior. When their food supply is exhausted, large numbers of single-celled amoebas of this organism coalesce into a wandering multicellular slug-like creature that then differentiates into an immobile spore-producing "fruiting body." The fruiting body has well-differentiated structures—a base, a stalk, and a reproductive head. In a sense, the cooperation of individual slime mold cells produces a coherent higher-scale individual as a slug and a fruiting body. The slug-like assemblies have a definite anterior and posterior, respond to environmental gradients, and have coordinated motility: that is, the complex internal interactions of the individual cells are hidden to produce a higher organization that has distinct interface with the external environment.

---

higher levels of organization, simulations and mathematics are important for understanding how neuronal circuits operate in behavior. Building and interpreting the results of mathematical models are important ways to gain insight into how the interactions of nonlinear processes give rise to system behavior.

Because biological systems are typically composed of a hierarchy of modular units, it is a challenge to gain an understanding of evolutionary dynamics at various levels of organization. But what accounts for the emergence of the modular units themselves: that is, what accounts for the evolution of the modular architecture? An example of evolving modular architecture can be seen in the genomes of organisms. Genomes are organized into modules at various different scales. At the largest scale, ensembles of genetic material are organized into chromosomes. Within a chromosome, material is organized into contiguous blocks of information that code for proteins, and sometimes groups of functionally similar proteins are organized into neighboring blocks called operons. However, there are variations of all kinds: introns that break up the protein-coding regions, alternatively spliced proteins, heterochromatin, euchromatin, dynamically remodeled chromatin modifications, insulators, DNA modification blocks, and introns within introns. Thus, there is a tremendous and dynamic variation in modularity of the genome within the same individual and across different species. What evolutionary forces govern the level of modularity in these genomes? Or more broadly, given the preponderance of modular organization in biological systems at all levels, what are the evolutionary dynamics that lead to such modularity? A typical modern computer central processing unit has 100 million transistors—it is impossible for a single person to design such a construct from the ground up with a global understanding of all the individual transistors. Thus, modular architecture allowing incremental buildup of complexity is an engineering imperative. Is that modular architecture also an imperative for biological systems? Does modularity have direct impact on fitness under some suitably posed dynamics? Is it a byproduct of selection for robustness or ability to evolve (such as the ability to generate variations)?

Network Topologies/Architecture

The interactions of modules form architectural organizations at a higher level. A common theme is that modules become hierarchically organized to produce modules at a larger scale. This characteristic is seen at multiple levels from metabolic pathways (Ravasz et al., 2002; Yu and Gerstein, 2006) to food webs (Pimm et al., 1991) and of course to the organization of interacting circuits in the brain. The recent explosion of functional genomics data has led to unprecedented large-scale assays of biological component interaction such as gene regulatory interaction and protein-protein interaction. A natural representation of such interactions is as a graph where each node represents an interacting unit (e.g., a protein) and each edge represents the functional interaction (e.g., physical collision). These graphs are commonly called networks, and the availability of large-scale networks has led investigators to notice certain statistical regularities in the structure of the node-edge connectivities (or the so-called *topology* of the graph). One statistical quality that has been suggested is that in many biological systems there are a few highly connected nodes while most other nodes are sparsely connected—this type of network has been called scale-free (Barabási and Albert, 1999; Jeong et al., 2000). It has been suggested that this statistical characteristic contributes to stable function in the face of network perturbation.

The representation of biological interactions as networks is a theoretical abstraction that has led to a wealth of descriptions of complex biological ensembles as architectural motifs. For example, network topologies seem frequently to have certain subnetworks that may allow certain kinds of dynamics or information processing (Yeger-Lotem et al., 2005; Jiang et al, 2006), and the network topologies may be correlated in functional (Magwene, 2001) and co-evolutionary groups (Qin et al., 2003; Tan et al., 2007). While a network is a static representation of component interactions, a dynamical view of biological processes may be obtained by considering how network topologies change over time. This approach has led to statistical characterization of dynamical structure of modularity in networks (Han et al., 2004) and, furthermore, suggestions that the logic of process regulation may be embedded in a dynamical network representation (Tu et al., 2005). The interest in network abstraction has been such that some have suggested that "network science" may comprise a new subdiscipline linking physics, biology, and chemistry and spanning scales from the molecular to the ecological (Barabási, 2002).

EMERGENT BEHAVIORS

The compass termites *Amitermes meridionalis* and *A. laurensis* build complex colony mounds reaching up to 20 feet in height with such distinct global properties as specific compass orientation and air columns that help regulate temperature (Korb and Linsenmair, 1999). These properties are thought to be ecological adaptations to local environments. The physical scale of the mounds is several orders of magnitude larger than the individual termites. The behavior of each individual gives no clue as to how the group manages to construct ventilation shafts that would seem to require a blueprint at the scale of the mound itself.

Closer to home, despite more than 50 years of intensive study, no explanatory model connects the activity of individual synapses to how humans store the memory of, for example, the face of an individual and then retrieve the same image among thousands of similar stored images (Kandel, 2001). As complex as are memory encoding and retrieval, they are simpler processes than higher cognitive processes like writing a sonnet. Thus, although biological systems are comprised of modules that hide the internal complexities, the composition of the modules generates unexpected ensemble behavior at a higher scale that is difficult to predict from knowledge of the parts themselves, even when the composition is of multiples of the same modules (*cf.*, termites within termite colonies). Because the collective behavior of these parts can be so surprising, the term "emergence" has been used to describe phenomena that seem to defy reductive understanding. This term has become somewhat burdened because of its use by some scientists to argue that certain natural phenomena cannot be understood by current scientific methods—a contention that is widely disputed. Nevertheless, the term carries an important metaphor of ensemble properties that are difficult to predict from our current models and therefore the term is used here in a strictly descriptive sense.

A reasonable way of thinking about emergent behavior might be to focus on the level or scale at which the rules reside. If the rules are specified at a low level, for example, the individual termites, and the patterns and structures, like termite mounds, emerge at a scale where there are no rules specified, we may call this emergent behavior.

Ideas of how some component interactions might give rise to emergent behaviors in biological systems can be deduced by analogy with engineered systems such as electrical circuits. For example, positive feedback loops, where an enzyme converted from an inactive to an active state in turn activates more copies of the same enzyme, are able to amplify small signals and give rise to large-scale switch-like responses to important but small changes in a cell's environment (Hlavacek et al., 2006). Conversely, negative feedback loops can dampen the effects of fluctuations in system inputs and allow cells to ignore uninformative noise in their environments. Combining feedback loops with time delays can give rise to oscillatory behaviors, such as the daily changes in plant metabolism that accompany the rising and setting of the sun. In many biological systems, relatively simple feedback loops and oscillators seem to act as modules within very large-scale networks. Some of these networks produce extremely stable overall behavior, such as the physiological regulatory mechanisms that maintain our body temperature and blood pH within very narrow ranges. Other networks are able to generate irreversible switch-like behaviors, as when different cell types within a developing multicellular organism become committed to particular cell fates (*cf.* Alon, 2006)(see Figures 7-1 and 7-3 in the next chapter). Many of the most interesting biological networks combine aspects of reversible and irreversible behaviors that in ensemble produce complex behavior, such as the ability of an animal's nervous system to learn and remember.

Earlier discussions in this chapter suggested that modules are made up of parts that collectively show coherent invariant properties in their interface with the exterior. These invariant external properties are derived from the complex and dynamic interactions of the parts and encapsulate the parts in a simpler and uniform interface with the environment. (Here the term "environment" refers to all that is external to an individual module, including other similar modules.) The "invariant properties for external interface" can also (again loosely) be called emergent properties of the ensemble that comprise the module. In the example of a slime mold slug, the ensemble displays an emergent property of coherent directional motion (Box 6-2). A protein is composed of a string of amino acids that, when placed in a solvent medium, folds into geometrical shapes. The folded structure then displays an emergent property of catalyzing chemical reactions with exquisite single-molecule specificity. Can emergent properties be predicted from the knowledge of parts? If so, would the same theory apply to all different scales at which modules can be identified?

Consider the problem of protein folding and prediction of its function. At first glance it appears that the only barrier to understanding is computational. A sufficiently fast and large computer could allow models of molecular motion in a force field to yield a prediction of equilibrium form. Given the form, the model of geometric lock-and-key for protein-based catalysis can be applied and brute force computations can be applied again to ask what reactions the folded protein would carry out. Then, are emergent properties simply those consequences of group interactions that are currently too difficult to compute? In the past 20 years, considerable development has taken place in the somewhat diffuse area of multiscale computing (Theodoropoulos et al., 2000; Brandt, 2001; Kobayashi et al., 2001). Those efforts cover areas from multiscale functional analysis such as wavelet analysis in signal processing and image analysis, multiscale clustering for databases, coarse-fine time-stepper dynamical systems (with recursion), multiscale optimizations, and piece-wise linear hybrid systems, as well as techniques such as Monte Carlo, Markov Chain, and grid computing. Those developments have enabled progress in many areas, including large-scale solid-state physics, fluid mechanics, molecular dynamics, image handling, genomics, and others. One idea common to those techniques is computation on small patches at a lower scale (microscopic scale) that can be used to interpolate at a coarser scale (macroscopic scale) but in a controlled, bounded manner. Bounded approximation at multiscales—as canonically expressed by the wavelet analysis—is an integrative approach that can be used to connect phenomena at different scales. A useful conceptual development in any area must be eventually connected to data and theory posed in a computable form. Development of multiscale computing and multiscale integration methods is critical to many of the cross-cutting questions discussed in this report.

Many of what are called emergent properties involve physical geometric form and direct interaction mediated by spatial and structural contexts. But many aspects of emergent properties, such as the construction of air shafts in termite mounds, require information processing among the participating components so that each component reacts in accordance with the information acquired from other components. Earlier in this chapter, cooperative interaction was suggested as a key ingredient in module formation. Bacteria, for example, form complex biofilms on human teeth. These biofilms contain hundreds of species in relatively stable interacting communities. Microbiologists call these interacting communities in which functions are divided "consortia." Many bacteria within these consortia make "auto-inducers," which are chemicals thought to permit communication, not only within, but also among, species. The existence of message-sending and message-receiving capacity in diverse species of bacteria has been facilitated by the ability of these organisms to exchange genetic information through the transfer of plasmids and phages. Therefore, a key ingredient for interacting parts, such as different species of bacteria, to display emergent behavior seems to be the existence of some process to pass information, sense information, and react to information. Information could exist as a minute quantum to a single part, but the collective computational action of the ensemble could lead to changes in synaptic strength that is the substrate

---

Box 6-3
Reconfigurable Robots as an Analogy of Emergent Properties of Biological Systems

The potential utility of understanding how communication and interaction of modular parts lead to emergent properties can be found in the area of engineering reconfigurable robots. Rather than engineering specialized robots for each specific purpose, it would be desirable to develop modular components so that robots could be reconfigured for specific tasks. This goal raises a number of interesting information problems. Assuming that a human programmer could provide the proper programming for each possible configuration of modules and that the modules are assembled in the proper form, how does the robot recognize its configuration and find the right program among the suite of available programs? Having defined the program (e.g., the robot is able to determine that the current configuration of modules is appropriate for detecting land mines, not evacuating wounded soldiers or entering a building), how does each module recognize what part it is to play in this program? That is, if there is a particular control process for the left side of the robot and another for the right side of the robot, how does each module recognize that it is in fact a module on a particular side? In a both abstract and very real sense, individual termites appear to carry out just such a computational paradigm and calculation to produce a mound. What is this computational architecture? Is there a common information-processing framework for emergent properties from individual synapses in a human brain to cells in a flowering plant to individuals in a community—an architecture that could be applied in systems engineered by humans?

---

of memory formation or the construction of a ventilation shaft in a termite mound. How components interact locally to produce global patterns is at the heart of the matter of emergent properties. A fundamental problem is what type of information transfer is carried out and how individuals are equipped to sense the input and act on it appropriately to produce the global patterns (Box 6-3).

In engineered systems, a basic step in creating complex behavior involves the construction of a broad implementation plan or architecture for the desired system. For example, the idea of computers communicating over shared lines with electromagnetic signals is a simple concept. The physical implementation of this idea, however, requires decisions on whether to encode bits in voltage or frequencies of the electrical signals, how different machines should share the same physical line, and a scheme to parse up the individual messages, just to name a few.

In biological systems, evolution has preserved a number of architectures that underlie complex processes. These architectures represent the broad implementation plans of an ensemble of components to produce emergent properties: that is, the Bauplan of the organisms at all scales (*cf.*, Raff, 1996). Details of the architecture determine the efficiency of the implementation, constraints arising from the chosen architecture, and all the possible different forms that can be derived from that architecture. Architectural considerations explain certain component actions that might be difficult to understand without the broader overview. For example, the presence of certain cells in the limb buds of mammals is difficult to understand unless it is known that the architecture of the developmental process calls for digit formation by programmed cell death of interdigit tissue rather than apical growth of digit tissue. Therefore, to achieve a conceptual understanding of emergent properties requires the development of a theory on the architecture of biological systems, a theory of the Bauplan applicable to scales from protein structure to ecosystems.

ROBUSTNESS OF BIOLOGICAL PHENOMENA

By technological standards, all organisms are highly complex, consisting of hundreds of thousands of interacting chemical species and thousands of regulated genetic elements. Intuitively, complexity seems to imply instability: The more things that can go wrong, the more likely the system will fail. Yet biological systems are stable. The ability of biological systems to maintain similar states or robust processes even when perturbed is manifested at all levels of organization from the regenerative dynamics of forests after a fire (so-called gap dynamics), to the development of whole organisms from fractions of the initial embryo (twins), to the stable folding of large proteins at boiling temperatures (e.g., within thermophilic microorganisms). Such robustness or stability is difficult to achieve in engineering settings; for example, despite many safeguards and redundancies, a single power station failure brought down the entire northeastern U.S. power grid in 2003.

It is hard to find any biological processes that do not have specific features that promote robust function under varying conditions. Developmental biologist C. H. Waddington coined the term "canalization" to describe organisms' ability to carry out the same function in various different environments. "Different environments" or "varying conditions" can also include genomic variability. Biological robustness can be classified into at least two types: robustness vis-à-vis environmental perturbations and robustness vis-à-vis genetic perturbations. Stable development of an embryo despite temperature fluctuations is an example of environmental robustness, whereas the redundancy of the genetic code is an example of robustness against mutational changes of DNA. A tension arises, though, when one considers that if an organism were completely robust to genetic change (in other words, if mutation never led to change in form or function), there would be no phenotypic variation for natural selection to act upon.

The ability to function despite external change, by contrast, seems to be an immediate fitness-enhancing factor, so a model to predict the evolution of robustness to environmental fluctuation might not be as problematic as robustness to genetic change (Wagner et al., 1997). Indeed as noted, homeostasis, that is, constancy in relation to perturbations, is a fundamental property of living systems. Are there general principles by which an organism maintains such robustness under a wide range of perturbations? What are the limits to that robustness? Why are organisms not omnipotent in their biological function?

Organisms clearly display robustness to environmental and genetic perturbations, but they also display a certain kind of "process robustness" that is related but not necessarily connected to robustness against perturbations. Recently, studies of molecular processes at the microscopic level (such as the cellular level) suggest that these molecular activities are extremely variable—or noisy (Samoilov et al., 2006). For example, with respect to gene transcription, a gene was considered as "on" or "off" or perhaps "highly expressed," but detailed measurements suggest that the transcriptional activity of an individual gene is variable and carried out in stochastic bursts (Elowitz et al., 2002). At the other end of the scale, simple ecological population dynamics such as prey-predator dynamics can be described reasonably well as a limit cycle, but the individual dynamics of prey capture, birth, death, and other parameters are extremely variable (Ellner and Turchin, 1995; Grenfell et al., 1998). Thus, dynamic biological processes, such as those in cell cycles, organismal development, or ecosystem nutrient cycling, may have very high component-wise variation. How, then, do these systems achieve precise system-level function despite such noise?

An important aspect of robustness is that biological systems display robust ensemble behavior at one scale despite the dynamic turnover of modules at a lower scale. An individual displays robust function while its component cells are constantly undergoing birth and death. Individual neurons maintain relatively constant activity patterns for much of the lifetime of the animal, despite the fact that the ion channels and receptors that control excitability are constantly being replaced at time scales of hours or days. Communities show consistent properties while individuals undergo birth and death, and even when an entire species becomes extinct. As discussed above, biological systems show robust external properties independent of internal complexities like turnover and noise. Thus, biological robustness is not just a static property obtained from materials or construction, but often a dynamic property where system function is maintained by dynamic organization such as various feedback and feed forward circuits or stable attractors.

Despite their robustness, biological systems also have profound vulnerabilities; for example, a single genetic change can cause a fly to develop without any eyes or with legs where its antennae ought to be (Raff, 1996). The fact that biological networks such as gene regulation networks have such critical nodes allows relatively small changes generated by mutation to give rise to new emergent properties. Within a given network, scientists currently lack general methods to predict which nodes are most likely to be critical and are, therefore, loci of both vulnerability and evolutionary opportunity and which nodes are relatively unimportant. A critical set of questions related to robustness is when or how a biological systems (e.g., networks of interaction) are robust and when or how they are sensitive (Samoilov et al., 2006).

Engineering robustness in manufactured products is an extremely difficult task. Using redundant parts is one standard engineering solution to increase robustness, but this strategy only works for catastrophic failures that can be recovered by backup parts—not for constant noise. Standard feedback control strategies also are applicable but only up to a certain degree of noise. By contrast, living organisms seem to have built-in mechanisms of robustness; remarkably, these robustness properties are distributed throughout different scales of organization. For example, because of the way that multiple codons can indicate the same amino acid, a protein is robust to some (but not all) possible mutational changes. When the protein folds, it is guarded against misfolding by chaperone proteins. If it misfolds, it is discarded by a quality control monitoring system and a fresh new copy of the protein is generated in its place. Even when a protein is entirely removed—for example, by deletion of a gene—the cell often has checkpoints for detecting such events and system-level regulation to compensate for the perturbation. Loss of cells in an individual can trigger stem cell proliferation. Likewise, aberrant cell proliferation is checked by induced cell death. At a different scale, individual loss from a population leads to increased birth rates and the loss of a particular group of species from communities—for example, loss of certain trees from wind damage in a forest—leads to adaptive recovery by other opportunistic species. Conversely, organisms are also exquisitely vulnerable to particular perturbations. Loss of a key predator could lead to a qualitative reorganization of a community, changing the level of a key molecule could lead to a switch in metabolism from dormancy to active proliferation, and changing the right set of amino acids could change the structure of a protein from an alpha-helix to a fundamentally different beta-sheet. Therefore, a key conceptual challenge is to understand and develop a theory of how robustness is promoted in biological systems and how it interplays with the control of sensitivity of the same systems. Formal mathematical models and simulations can quantitatively explore the boundaries of parameters

for robust action, and this kind of question is one for which formal models and simulations are useful.

Although there is encouraging progress in demonstrating robustness and its evolution, there are still great challenges ahead and unforeseen implications. One possibility is that interplay between robustness and breakdown of robustness can facilitate the evolution of novel phenotypes through the accumulation of hidden genetic variation—thus, robustness may be precisely the characteristic that produces unexpected new forms. Another possible consequence of robustness came from a simulation study on the evolution of RNA secondary structure (Ancel and Fontana, 2000). That study showed that evolution of robustness in RNA secondary structure led to a selection for modular decomposition of different "morphological" elements (stem loop regions) in the melting profile of the molecule. Those results suggest that modularity could be a coupled feature of robust systems. There are still challenges to understanding the interaction of modular architectures and robustness properties, and using model systems and high-performance computational simulations will be an important approach.

Studies of molecular processes at the microscopic level (such as the cellular level) suggest that molecular activities are extremely variable—or noisy (e.g., Pedraza and van Oudenaarden, 2005; Rosenfeld et al., 2005). At the other end of the scale, individual turnover in a food web is subject to great variability and stochasticity. How do these processes with high inherent variability achieve precise system-level function despite such noise? Synthetic gene regulatory circuits have been constructed in bacteria that display precise dynamic behavior such as an inducible bi-stable switch (Isaacs et al., 2003). However, the dynamic behavior is usually at the mean population level and individual cells vary widely in their dynamics. If this were generally true, how would a multicellular organism ever function? In fact, how would even a single cell function when all of its processes could end up being uncoordinated? Similarly, evidence shows that a yeast cell may contain only a few copies of many of its RNA transcripts. Any regulatory processes involving these transcripts—unless mechanisms are in place for precise single molecular reactions—are likely to be stochastic; mass kinetic models as used in standard chemistry cannot apply to these molecules. Could the dynamic principles of control processes in organisms be completely different from standard systems such that they allow inherently robust dynamics from noisy components (Samoilov et al., 2006)? In fact, could noisy dynamics be an adaptive characteristic as suggested for individual cell behavior in bacterial chemotaxis (Korobkova et al., 2004)? Coupled chaos as a model for quasi-stable ecological communities has been suggested; could such control processes operate at all scales of biological systems? Many models of biological processes in cells, organisms, or even ecosystems are derived from static and coarsely quantitative measurements.

Once real-time in vivo dynamic measurements of large-scale multiple components are achieved, biologists will likely develop a different view of biological control processes and the resulting robust functions. Such data will likely demand very different theoretical models of biological function.

CONCLUSION

The question "What are the engineering principles of life?" begs the development of a conceptual framework for understanding how biological systems take on particular forms and robustly carry out their functions. Even a small part of an answer to this question, especially when derived in a computable form applicable to data, will have great impact in all subfields of the biological sciences. For example, in protein and metabolic engineering, even an approximate understanding of how to manipulate modules to produce desired forms or biochemical processes would be highly desirable. The field of biomimetics attempts to use biological engineering principles to generate devices that have desirable biological properties such as robustness and reconfiguration. Restoration ecology attempts to manipulate certain community ecological functions to remedy human perturbations to ecosystems—at the largest scale, even up to possible remediation of the effects of global warming. All of these applications require computable predictions of emergent properties and understanding of how biological systems achieve robustness. At a grand level, understanding human cognitive function requires an understanding of how modular processes from individual synaptic vesicles, to synaptic boutons, to neural networks, to neuroanatomical regions all integrate across scales to enable speech, memory, and thought.

Is modular architecture a necessary requirement for generating complex biological objects? Modular construction is a human engineering concept and need not be a characteristic of evolutionarily derived biological objects or ecological assemblies. Does the process of evolution promote the appearance of modular units? If so, are certain architectural elaborations likely or inevitable? Understanding the emergence of modular architecture across all scales and its possible contribution to properties unique to living systems, such as variation and robustness, is a key conceptual challenge of the future.

Twentieth century theoretical biology provided the framework for mathematical and probabilistic dynamics of the turnover of individual components (such as alleles) in a closed system (such as population). In a way the theoretical foundations are similar to that of classical mechanics in physics. Given a closed system, the theory makes predictions about the motions of indivisible units. Similar to the development from classical mechanics to solid-state physics, pattern formation, and engineering, in the 21st century, development of a new theoretical framework is necessary for understanding how the ensemble of units produces new emergent modules and emergent properties and for understanding the architectural principles of how biological systems are assembled from such modules across scales from individual molecules to entire ecosystems.

As the new theoretical framework develops, constant evaluation and reevaluation are necessary to evaluate whether any given theory can make "predictions." In physical systems, some processes are intrinsically unpredictable because they involve features that can only be described in probabilistic terms. Other systems are unpredictable even though they are completely deterministic, because of their chaotic dynamics and extreme sensitivity to initial conditions. For some physical systems, attempts to predict fail simply because the important controlling details of the components are not well understood. Most current biological modeling implicitly assumes that accurate predictions can be made once sufficient information about the biological system is available. However, it is possible that some biological processes will be intrinsically unpredictable because of principles analogous to chaos or quantum indeterminism. A fundamental goal of chemistry or engineering is to understand and predict the behavior of compositions of parts. A major area of biological theory will be developing a similar understanding of constructive principles of biological organizations.

7

What Is the Information That Defines and Sustains Life?

Advances in technology have dramatically increased the amount of biological data being collected. For example, DNA sequence data on the genes of many organisms and satellite imagery of many ecosystems are now available. While this proliferation of data presents exciting new opportunities, making good use of it also presents significant challenges. Increasingly powerful computing technology provides a powerful tool for data analysis and allows for the use of such techniques as shotgun sequencing of whole genomes. However, in many cases using data to come to meaningful conclusions about life requires time-consuming and expensive work. The sharing of data sets between researchers provides opportunities to examine data collected for one purpose to make progress in a different area. For this type of sharing to be most productive, data sets need to be codified, well curated, and well maintained in an accessible format. While data sets are essentially collections of information, the role of information in biology is much more than the use of data sets. The refinement and application of theories of information to biology present a deep challenge and an opportunity for furthering our understanding of life. Existing theories of information borrowed from other fields can be difficult to apply to biology, a field in which context is so important, but the conceptual gain may be well worth the challenge.

WHAT IS INFORMATION?

The concept of information is used throughout biology. Biologists study how information is acquired, used, stored, and transferred in living things.

Many biological structures or processes can be thought of as carriers of information. From the sequence of a DNA molecule, to sounds, nerve impulses, signaling molecules, or chemical gradients, scientists find it useful to characterize biology in terms of information. From the critical discovery of the "genetic code" as the coupler between DNA sequence and protein synthesis, to the marvelous ability of bees to convey information about the location and quality of resources through dance, it is intuitively appealing to describe the processes and structures of biology in information terms. Throughout this report, there are numerous examples of the representation and transmission of information. Questions that naturally arise about information in biology include these: Is there a common way to think of the biological information in all of these representations? Is there a consistent and useful way to measure biological information so that it can be dealt with in quantitative descriptions of genetics, evolution, molecular processes, and communication between organisms?

In common usage, the word "information" conveys many different notions. It is often used as a synonym for "data" or knowledge, and in most common language uses it is associated with written or spoken numbers or words. This connection is key to a more scientific use of the term, in that it suggests that information can be represented by numbers or letters or more generally by *symbols* of any form. Indeed, information must have a representation, whether it is as written symbols, bits in computers, or in macromolecules, cells, sounds, or electrical impulses. The informal use of informational terms is widespread in molecular biology. For example, molecular biology uses words that relate to transfer and processing of information as technical terms for biological processes. The choice of words like code, translation, transcription, messenger, editing, and proofreading reflects how scientists think of these processes. When information is used as the focal concept for thinking about molecular biology, it highlights the sequence properties of the molecules under study, instead of their actual physiochemical forms (Godfrey-Smith, 2007). This prompts a focus on the abstract representational role of these molecules, rather than the nature of the physical processes (e.g., the biochemistry of the translational machinery) that are inevitably required to express the stored information in meaningful form. It is important to think carefully about information at many levels, both below the sequence level in molecular detail and above at higher levels of organization.

INFORMATION IN BIOLOGY

August Weismann appears to have been the first to explicitly use the notion of information transmission in genetics in 1904 when he referred to the transmission of information in heredity (Weismann, 1904). The metaphor of information represented in written or electronic forms has become widespread, but is it more than a metaphor or an analogy? The concept can be formulated quantitatively for molecular biology, perhaps most strikingly in the operation of the "genetic code" physically embodied in the form of DNA and RNA. In the genetic code, sequences of three nucleotide bases, called codons, are used as symbols for the amino acids in proteins—and the so-called translational machinery of the cell biochemically synthesizes proteins from the coded instructions represented in RNA. In this case, the amount of information transmitted, for example, can be calculated as it goes from the DNA to the RNA and then to the protein, in which process a fraction of the information (about half) is typically unused or lost (Dewey, 1996, 1997).

Biological systems differ from nonliving systems in several ways, but the most profound differences might lie in their information content. It can be useful for this purpose to think of biological systems as evolved transducers of information, since organisms accumulate, process, store, and share information of different types and on different time scales. An organism needs information about its internal condition to manage its internal functions. For example, organisms use internal information to maintain homeostasis, to coordinate and regulate development, and to detect potential pathogens. (See Chapter 8 for more examples.) Information about the external world also is critical for an organism to deal effectively with that world—for example, organisms use external information to find shelter, to escape from predators and compete with rivals, and to reproduce and care for offspring. Information about the structure and physical function of the organism also is necessary for evolution to proceed, and this information is sequestered in the genome in a variety of ways, some of which are not yet understood. The information described above is represented in a variety of forms, probably none more well known than the digital information in the genomes of organisms. This information is central to biology, for it represents the largest share of information that is passed on in reproduction (Hood and Galas, 2003).

The field of genetics investigates the way in which symbolic information in the genetic material is inherited and interpreted as messages about protein and RNA structure or as messages about the timing and levels of gene expression. Cell biology seeks to understand how intracellular components encode and interpret the information necessary to organize cellular structure, maintain homeostasis, and carry out cellular functions. Development can be seen as the study of how these messages are used to extract and interpret the information in the genome in order to turn a single cell into a complex multicellular organism composed of thousands of cells with specialized functions. Neurobiology is the study of how internally generated electrical signals are combined with information about the environment to allow the animal to generate meaningful behavior. Immunology depends critically on the problem of detection and surveillance—that of distinguishing self from harmful invaders such as bacteria and viruses or aberrant self such as cancerous cells. This detection problem requires exquisite sensitivity and precision, as does the process of mounting and regulating the appropriate responses.

The storage and transmission of information are fundamental to living things, but they are not the exclusive properties of life. For inanimate matter, the power of information storage and transmission is decidedly limited, but not entirely absent—for example, crystals, dendritic minerals, snowflakes, and other physical and chemical structures form, and thereby store, information in spontaneous order. In living things, however, the power of information acquisition and transmission is enormous, characteristic, and almost unlimited in potential. The transition from the inanimate to animate might well be thought of as the acquisition of the singular ability to increase the storage and transmission of information, in quantity and quality. The possibility of this increase of information, well beyond what is ever seen in inanimate matter, is fundamental to the process called evolution. Darwin's marvelous ideas, which embody this concept in a qualitative fashion, can be viewed as the realization that variation and selection are the key characteristics of this potential and that they interact to accumulate information in living lineages. The idea of evolution has, in fact, been recast in modern times in terms of information flow. Evolutionary biology can be thought of as the study of how information enters the genome, persists, and changes over time—the ebb and flow of information, its gain and loss.

Developing a conceptual treatment of information measurement, storage, and transmission in biology will require logical discipline. The process of doing so will elucidate—and raise new—questions about the dynamics of evolution and the processes of physiology, development, and behavior and perhaps will even shed light on the origins and the fate of living systems.

INFORMATION THEORY

While the concept of information in biology makes sense using a common-language perspective on the term "information," and while it captures the symbolic or representational nature of much biological information (Godfrey-Smith, 2000), an adequate definition of the term "information" for formal use in biology remains somewhat elusive. The two fundamental questions are: First, how is a particular kind of biological information represented or encoded? And second, how can the quantity of information in a given representation, biological or not, be usefully defined and, most importantly, how can it be measured?

Some guidance is available from the large body of theory and research

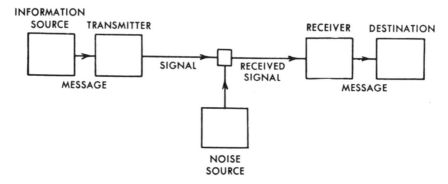

FIGURE 7.1 Shannon's framework for thinking about information transmission. Information from a source is encoded or represented by a transmitter, which sends that information through a (possibly noisy) communication channel in the form of a signal. The signal is received by a receiver that decodes the message and delivers it to the destination. C.E. Shannon, (1948). From The Mathematical Theory of Communication. Copyright 1949, 1998 by Board of Trustees of the University of Illinois. Used with permission of the University of Illinois Press.

that deals with information outside of the field of biology. One approach is provided by information theory as founded by Claude Shannon and Norbert Wiener to quantitatively understand communication channels (Shannon, 1948; Wiener, 1948; Shannon and Weaver, 1949) (Figure 7-1).

In Shannon's theory, information is essentially that which allows its bearer to distinguish among alternative possibilities—the range of possible messages. By this approach, the more alternatives that can be distinguished among, the more information has been transmitted (Box 7-1). For example, sitting in a windowless office, one cannot distinguish among different possible weather conditions outside. It might be sunny or cloudy. By checking the weather on the Web, one can find out which of these states is actually occurring. Thus, the weather report contains information. If the weather report also gives the temperature, then one can distinguish between even more states: sunny and hot, sunny and cold, cloudy and hot, cloudy and cold. In this case, the report provides more information than if it only indicated current cloud cover. This view of information is closely related to notions of communication and computation; the amount of information conveyed by a signal is proportional to the bandwidth that would be required to send that signal through a communication channel or the storage space that would be required to record the message in compressed form on a computer.

In the Shannon approach to information, the message is the result of a transmitting source sending a signal in some given representation, usually a symbolic alphabet. Rather than assigning an information content to any specific message, the amount of information sent through the channel depends only upon the characteristics of that source and the range of possible messages that might be sent.

Neuroscientists have an advantage in adapting formal information theory to their work, as spike trains can be easily understood to carry information about sensory inputs. However, in other fields of biology, it can be more difficult to define the properties of the "source" and the symbolic alphabet used in representing biological information in any satisfying way, so this approach presents conceptual problems. For example, to calculate the amount of information in the amino acid sequence of a protein requires knowing how many such sequences are possible. The question then is: What does that mean—literally all possible amino acid sequences of that length, or all possible sequences represented in living organisms, or all possible sequences in the currently known database of protein sequences, or some other way of characterizing the possibilities? These different possible "sources" would all yield different measures for information. This is clearly a problematic approach.

Algorithmic Information Theory

An alternative approach to defining information brings out the role of information in computation. Rather than measuring the information content of a statistical source, as Shannon does, algorithmic information theory considers only the message itself and asks what is required to generate or reconstruct just that message. This inherent "complexity" idea comes from a formulation known as Kolmogorov complexity, after the Russian mathematician Andrei Kolmogorov. The key concept was independently arrived at during the 1960s by Ray Solomonoff (1964), Geoffrey Chaitin (1966), and Kolmogorov (1965). This simple but subtle idea holds considerable promise for biology. It is currently heavily used in imaging processing, pattern recognition, artificial intelligence techniques, and other engineering applications, but it is just now beginning to be used in biological applications. For example, this powerful approach has been applied to calculating mitochondrial genome phylogeny (Li et al., 2004).

In addition to the difficulties discussed above, it is becoming clear that most biological information depends on the context in which it finds itself—what other information is present in the same system, and how that information influences the range of actions that a protein, cell, or organism can take. If the representation of information cannot be "read" or used when it is out of context, it carries no meaningful information. For

Box 7.1
The Mathematical Basis of Shannon's Ideas

Suppose we receive a message *m* that could take any of four possible forms, A, B, C, or D, each with equal probability. How much information is associated with message *m*? Because the message allows us to distinguish among four different alternatives (A, B, C, or D), we might be tempted to say that *m* conveys four units of information. But suppose that we receive two such messages, *m*1 and *m*2, one after the other. Intuitively, it would be nice to say that this pair of messages gives us twice as much information as did the single message *m*. But notice that this pair of messages actually allows us to distinguish among not eight but rather 16 equally likely possibilities. By doubling the number of messages, we have quadrupled the number of alternatives among which we can distinguish:

AA	BA	CA	DA
AB	BB	CD	DB
AC	BC	CC	DC
AD	BD	CD	DD

In a series of early (1917-1928) papers, Harry Nyquist and R. V. L. Hartley pointed out that if we measure information by the *logarithm* of the number of alternatives that can be distinguished, the problem is resolved. The message *m* gives us log(4) units of information. The pair of messages *m*1 and *m*2 together give us log(16) = 2 log (4) units of information—exactly twice what we obtained from the single message alone.

Now what happens if the different messages have different probabilities of occurring? Suppose that message A is sent with probability 7/10, while messages B, C, and D occur with probability 1/10 each. In this situation, it seems that if message B comes through, we've learned more than if message A comes through. Each message—A through D—allows us to distinguish among four alternatives, but somehow we seem to have learned more when we receive message B than when we receive message A. After all, in the absence of prior knowledge we would have been "expecting" the signal A anyway, so when A does arrive this doesn't come as a particular surprise. Can we capture this somehow in our definition of information?

Consider another example. Suppose there are 10 possible states of the world: A1, A2, A3, A4, A5, A6, A7, B, C, and D. Then if we receive signal B, this

example, a pheromone or vocal call of one species commonly conveys little information to another species (or at least a very different kind of information); a common human gene, rich in information for a human cell, is likely to carry no meaningful information in a bacterial cell; a segment of amino acid sequence that folds into a functional protein structure in the context of the sequence of its native protein may be useless and nonfunctional when set in the context of another protein sequence; the structure of an orchid's allows us to distinguish among 10 states of the world. Signals C and D are the same; each of these provides us with log (10) units of information.

If A1 occurs, this also has information log (10), but if we simply receive the signal A in response to this event, we actually don't find out whether A1, A2, A3, A4, A5, A6, or A7 actually occurred. Thus we have lost the ability to distinguish among seven alternatives; the net amount of information that we get is then

$$\text{Log (10)} - \text{Log (7)} = \text{Log (10/7)}$$

This suggests a measure of the information provided by a signal S that occurs with probability p:

$$\text{Information(S)} = -\text{Log } p$$

Applying this to our example above:

$$\text{Information(A)} = -\text{Log } 7/10 = \text{Log } 10/7$$
$$\text{Information(B)} = -\text{Log } 1/10 = \text{Log } 10$$

At last we are in a position to define the expected amount of information transmitted by a signal. Suppose that, as in our previous example, the message m takes the form of one of four signals, A, B, C, and D, with probabilities 7/10, 1/10, 1/10, and 1/10, respectively.

Then with probability 7/10 we will get a signal (A) that provides Log 10/7 units of information, and with probability 3/10 we will get one of the three signals (B, C, or D) that provides Log 10 units of information. The average, or expected, amount of information provided is then 7/10 Log (10/7) + 1/10 Log (10) + 1/10 Log (10) + 1/10 Log (10).

More generally, we can say that if symbols i = 1, 2, 3, . . . , m, are sent with probabilities p_1, p_2, p_3, ... pm the average amount of information H in a message is given by

$$H(p) = -\sum_{i=1}^{m} p_i Log(p_i)$$

flower might facilitate pollination only for a single species of insect, and so on. Almost all biological examples have some contextual content. The information measures discussed earlier do not explicitly take context into account, as they were purposefully designed to be context-free. For biology, context is almost always essential, and consistent and useful theoretical tools are needed to describe, measure, and use contextual information in complex biological systems.

Decision Theory

The field of decision theory more directly accounts for context when measuring information. This is a body of theory that is designed to study optimal choice behavior. In decision theory, information allows its bearer to make good choices in an uncertain world. Information is measured not by the bandwidth required to convey it, or its statistical structure, but rather by its value. The value of information is measured by the best payoff one expects to get from a decision based upon that information, minus the best payoff one can expect if one has to make the decision without that information. For example, an investor can gain higher expected returns from the stock market if she knows more about the corporations in which she invests. Information about these corporations is measured by the difference in expected returns. These ideas have found fertile ground in application to biological information problems, particularly in evolutionary ecology and behavioral biology. There, in decision problems and game-theoretic scenarios alike, information is routinely measured by its influence on expected fitness.

Evolution establishes a relationship between the quantity of information and its usefulness, but whether this relationship is general, specific, or even expressible in a succinct form is not known at the moment. The need for more theory in this case is evident, but valid and precise information measures and probably a lot more data are necessary for the development of those theories. Then, perhaps biologists will be able to construct good quantitative theories that use information as a key measure in biological systems and begin to understand biological complexity in a quantitative, consistent, and useful sense.

STORING AND EXPRESSING INFORMATION IN THE GENES

The discovery of how biological systems transduce genetic information was one of the most profound triumphs of 20th century science. Somehow, the cells of an organism contain the hereditary information that—given appropriate interactions with the environment—determines phenotype and behavior. Over the course of a century, researchers in the field of genetics have largely worked out the common set of mechanisms by which all living organisms represent and express the hereditary information in their genes, leading to a detailed understanding of the mechanistic basis of heredity (see also Chapter 9).

Several questions remain to be answered, however, in order to fully understand how a system uses this information. First, the information must exist in some physical form; what is the chemical, mechanical, or electrical structure in which it is represented? Second, what does the information encode, and how are the details encoded and expressed? Third, how is the information in its physical form transduced so that it can be realized in phenotype and behavior?

In the case of the genetics, the rules of heredity of certain properties or traits of organisms, as discovered by Gregor Mendel, were rediscovered at the beginning of the 20th century. A major initial discovery was to determine exactly where this hereditary information lay. At the turn of the 20th century, Boveri, Sutton, and Morgan realized that the known rules of heredity could be explained if the heredity information was somehow contained in the chromosomes. Fifty years later, Hershey and Chase devised a stunningly simple experiment that used their knowledge of bacterial viruses and a kitchen blender to provide strong chemical and physical evidence that the DNA component of the chromosomes was the actual information carrier (Hershey and Chase, 1952).

Beadle and Tatum (1941) suggested that the information in genes describes how to make proteins: They postulated that genes affect function because each gene encodes a single protein. The next conceptual step was to figure out how the structure of DNA encodes information and how that information can determine the formation of a complex organism. In principle, this could happen in a number of ways. For example, DNA might form some kind of geometric templates for complex proteins. It might form some kind of polymeric substrate for driving thousands of different catalytic reactions. Or, as turns out to be true, DNA could be a coded instruction set that is read and decoded by another sort of molecular machinery. Watson and Crick inferred the rules that revealed the now famous structure of double-helical DNA molecules (Watson and Crick, 1953). Crick, Brenner, and colleagues figured out that there was a triplet code in the DNA so that each three base pairs of DNA determine one amino acid of the resulting protein (Crick et al., 1961). Subsequently, Nirenburg, Khorona, Holley, and others worked out the coding rules by which DNA sequences are subsequently translated into proteins primarily by using synthetic RNA molecules in biochemical reaction mixtures for making proteins in the test tube. These rules are now known as the "genetic code" even though it is now known that much more than the protein sequence information is contained in the DNA molecule of every organism.

Much of the subsequent revolution in molecular biology that unfolded in the last half of the 20th century elaborated biologists' understanding of how each step of this process works: how DNA encodes protein structure, how the cellular machinery translates this code into proteins, and how the rest of the molecule provides information for the control of which proteins to synthesize and when. Thus, a complete picture of DNA has been developed as a uniquely stable molecule that stores complex specifications for building and managing the organism. The specification can be found in a pattern-based code that depends on the linear sequence arrangement of its monomers (base pairs) rather than the DNA molecule's collective mechanical or chemical properties.

From this picture it can be concluded that the complex of reactions catalyzed by proteins can be orchestrated by how and when the information encoded in the DNA is expressed, but how the regulation of DNA expression impacts protein behavior was not at all obvious. The issue of how cells process information and make the computations that control protein expression became a major conceptual problem (Jacob and Monod, 1961). The revelation of the DNA structure and the genetic code opened the door to this problem; biologists now understand the control of gene expression to some extent, but its full complexity remains to be unraveled. Expression of information from the DNA into structural parts, catalytic enzymes, and other macromolecules drives a large part of the complex structures and functions of biological systems—cells, organs, organisms. Biologists are just beginning to figure out the patterns, the rules, and all of the machinery that generates this complexity from the information stored in the DNA. In fact, the theoretical underpinnings of this general problem—the conceptual basis of the global control of gene expression—is one of the major modern challenges of biology.

Jacob and Monod worked out how a bacterial cell controlled the expression of the set of genes it used to take advantage of a particular energy source (sugar lactose) that it encountered (Box 7-2). That work illustrates in a simple form how molecular machinery and the information processing of the bacterial cell informs us about the regulation of gene expression.

The basic components of the lactose operon are shown in Box 7-2 (part b). The lactose repressor is encoded in a nearby gene, the *lacI* gene. This protein is produced at the same low level all the time, independent of the medium or the metabolic state of the cell. It forms a tetramer of four identical units that recognizes and binds to a specific DNA segment that overlaps the promoter of the lactose metabolizing genes—when the repressor is bound, transcription is off. The turning on of the expression of the metabolizing genes of a particular substance is an example of what Monod and Jacob called induction. In this case, lactose is the inducer. The inducer binds directly to the lactose repressor and causes the protein itself to change its conformation, rendering it incapable of binding tightly to the operator. This is the basic induction response of the lactose operon—a disabling of a negative regulatory mechanism that allows transcription of the gene to proceed.

Despite its name, this operon is sensitive to factors other than the presence or absence of lactose. The cell does not need to metabolize lactose if other carbon and energy sources are available. Glucose is the preferred energy source in bacteria because it is a highly energy-efficient carbon source.

(It is one of the products of catabolism of lactose by β-galactosidase). The ability of glucose to regulate the expression of a range of operons ensures that bacteria will utilize glucose before any other carbon source as a source of energy. The ability of glucose to control the expression of a number of different inducible operons is accomplished through a protein called the cAMP-binding protein (CAP; Box 7-2b). A key observation in deciphering this mechanism was the inverse relationship between glucose levels and cAMP levels in *E. coli*. When glucose levels are high, cAMP levels are low; when glucose levels are low, cAMP levels are high. Biologists now know that this relationship exists because the transport of glucose into the cell directly inhibits the enzyme adenyl cyclase that produces cAMP. The cAMP then binds to CAP in the bacterial cell. The cAMP-CAP complex, but not free CAP protein, binds to a site on the DNA in the promoters of catabolite repression-sensitive operons. The binding of the complex enhances the activity of the promoter and thus more transcripts are initiated from that promoter, so that there is a positive control. The logic of this module of functional regulatory control then is the following. If there is little or no glucose present, and lactose is available, the operon turns on. There are two inputs and one output.

The lactose module can be thought of as an integrator of sorts. If the regulatory response were binary, or Boolean—on or off—it can be considered as an "AND gate." While the lactose operon is complex in the sense that several proteins, specific DNA protein interactions, induced conformation changes in the repressor and the CAP protein, metabolic sensors, and enzymatic activities are involved, it behaves like a simple "AND gate" as depicted in Box 7-2c from the point of view of the cellular logic. The quantitative aspects of the behavior of the operon are important for some aspects of the cell's response, so that the Boolean model is insufficient in detail, but the basic response is really very simple.

The lactose operon system processes information about the environment of the cell in order to regulate the expression of information stored in the genome of the bacterium. However, it is unclear how consistently it describes and measures the information, be it from the environment or from the genome. It is difficult to describe and measure the relevant information because of the complexity of the local environment, the diversity of information types present, and the complexity of the genome. Bacteria typically have a few thousand genes and a few million base pairs of DNA in their genomes, whereas mammals have 25,000 or so genes and a few billion base pairs of DNA in their genomes. Although comprehensive models of gene expression, particularly in simple bacteria and archaea, are being developed, biologists' understanding of the global regulation of gene expression in any multicellular organism is far from comprehensive (Bonneau et al., 2006).

Box 7-2
The Lactose Operon: A Genome-Encoded Network

In the 1950s, researchers noticed that the bacterium *E. coli* synthesizes the lactose-metabolizing enzyme β-galactosidase only when lactose is present in its growth medium. Jacob and Monod and their colleagues focused on this phenomenon and hypothesized the correct explanation of their observation. The explanation was elaborated into the "operon model," and the field of molecular gene regulation—still a major research area in biology today—began in earnest. The lactase genes (there are three rather than just the lactose-cleaving enzyme, β-galactosidase) are copied or transcribed as a single mRNA unit that is common for bacteria, and control is accomplished at the level of transcription of this single mRNA. The three structural genes that code for the protein enzymes involved in lactose metabolism are the *lacZ* gene that codes for β-galactosidase (β-galactosidase is an enzyme that breaks down lactose into glucose and galactose); the *lacY* gene that codes for a permease (the permease is involved in uptake of lactose from the medium into the cell); and the *lac A* gene that codes for a galactose transacetylase. These genes are transcribed from a common promoter into an mRNA, which is translated to yield the three distinct enzymes. Because the critical factors that the cell is responding to are metabolic in nature—the need to use lactose as a carbon source—the genetic regulatory network is coupled to the metabolic network of the cell. The lactose system was a fortunate choice by Jacob and Monod because it turns out to be a very simple system indeed—at least by the standards of genetic regulatory networks.

Structure and function of the lactose operon. (a.) The organization of the lactose operon is shown along the (blue) genomic DNA molecule. The promoters are shown as red arrows, and the regulatory sites on the lac promoter, the CAP-binding site and the repressor-binding site, or operator, are shown as blue and green boxes respectively. (b.) The regulatory flow of information is shown in blue (genetic regulation) and orange (metabolic), illustrating the essential components that are coupled across the boundary between the metabolic and genetic domains. (c.) The logical structure of the regulatory relationships is summarized in this diagram that uses symbols common in electronic logical operations. The simplicity of the basic logic is evident here even though the biochemical and genetic interactions underlying the logic are much more complex.

REPRESENTING INFORMATION IN DEVELOPMENT

In the process of development, information from the genome is used to execute a program of cell division and change (differentiation) to create a multicellular organism from a single cell. The early embryo develops from a single cell, driven by a network specified by information in the genome. This information comes in two forms: the DNA sequences that are binding

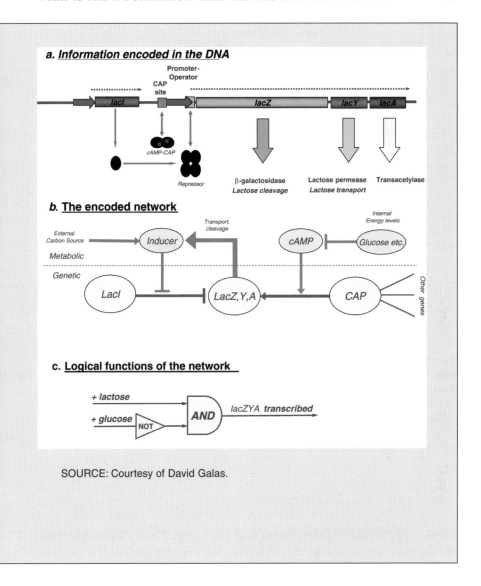

SOURCE: Courtesy of David Galas.

sites for a number of proteins in the DNA and the DNA sequences that are actual protein-encoding segments of DNA. These proteins bind to sites in the DNA near genes, some of which encode other proteins that bind DNA sites and regulate their expression (transcription factors). The structure of one such network is indicated in Figure 7-2.

Davidson and colleagues have mapped out this network for the sea urchin embryo (Bolouri and Davidson, 2003; Howard and Davidson, 2004;

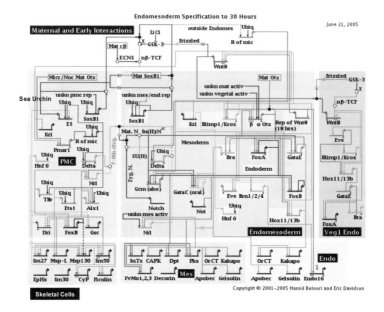

FIGURE 7-2 Sea urchin embryo development network. The structure of a network that executes the program of cell division and differentiation in the early sea urchin embryo. The genes are represented by the short horizontal lines; the control relationships are depicted by the lines extending between these genes. The different modules that control gene expression in different components of the early embryo are indicated in color: The pink segment is the skeletal cell module, the green segment is the mesoderm module, and so on.
SOURCE: Reproduced with permission of Eric H. Davidson.

Levine and Davidson, 2005; Istrail and Davidson, 2005). What this static picture does not show is the dynamics of the changing levels of gene expression as the program unfolds in time and that the first 30 hours of development of the embryo's life is driven by the dynamic network. There is also unseen complexity in the batteries of other genes, including the metabolic and structural genes that are expressed in each cell type driven by the presence of the specific set of transcription factors in the cells of each type. This example illustrates the nature of the information needed and the degree of complexity involved in early embryogenesis. In many ways, the most remarkable thing about this work and the resulting genetic regulatory network is that it is decipherable and understandable at all. The dynamics of the gene expression program that leads to the early sea urchin embryo is one of the most explicit cases known to date where the informational definition of the network has such clear biological significance. This advance will soon be only one of many such cases, and the specific and quantitative role of genetic information in embryogenesis will soon be much clearer.

Study of gene regulatory networks in different organisms suggests that various subroutines are employed repeatedly. As discussed in Chapter 6, these conserved "modules" provide a circuit that drives a particular kind of outcome. The kind of circuit needed for development differs from the circuits characteristic of physiological regulatory networks like the lac operon described above. Box 7-3 gives examples of a particular circuit used in several different developmental pathways.

SHARING INFORMATION

Much of the accumulation of biological complexity that has occurred over the history of life on Earth has arisen through major transitions in which previously unassociated entities either joined into a common reproductive fate or developed cooperative associations while maintaining reproductive independence (Maynard Smith and Szathmáry, 1995). These transitions have had a number of effects, such as economies of scale and functional specialization.

For example, symbioses that developed into complete cellular dependence occurred at least twice in the history of life: in the acquisition of mitochondria in eukaryotic cells and in the acquisition of chloroplasts in algae cells. It is significant that the incorporated bacterial cells that became mitochondria and chloroplasts retained part of their genomes and the ability to replicate them when they joined forces with their eukaryotic hosts. The cell acquired an entire new genome (or two). After incorporation and transfer of many genes into the nuclear genome, symbiosis between engulfed prokaryote and host eukaryote eventually evolved into a full partnership. Coordination of the symbiosis that led to the full partnership required the cellular genome to communicate with the organelle genome in ways that finally became permanently fixed in the information of their respective genomes. The evolution of mitochondria and chloroplasts is an illustration of how the complexity of biological information can increase. These intracellular organelles require cells to have a new level of communication and coordination.

Information sharing works differently than the sharing of physical resources in that it is not a "zero sum game," as expressed by the British playwright George Bernard Shaw:

Box 7-3
Comparison of Developmental and Physiological
Regulatory Networks

Unlike many physiological regulatory networks that have the purpose of moving the cell to a new state in response to the environment (see Box 7-2), developmental regulatory networks are more like sequential computer program subroutines in that they drive the unfolding of a defined set of successive steps or stages, as the program executes over time. Whereas the dynamical properties of the physiological networks enable them to transition back and forth between states in response to a changing environment, developmental networks, while sensing and coordinating with the cellular environment, must drive a regular, irreversible series of transitions through a defined series of states. Developmental programs, at least in the early embryo, probably never get near a steady state. The program inexorably drives itself forward, unfolding each successive stage of gene expression in the appropriate cell types. How do these kinds of programs work? Is there a theme or a repertoire of mechanisms?

A number of examples of network mechanisms that drive development systems forward are known. The regulatory interactions of a small set of genes that drive the transition of the network to the next stage are a recurring theme. Despite the variety of organisms and cell differentiation pathways represented in the four examples shown here (two from the sea urchin (a and d), one from the mouse (b), and one from the fruit fly (c)), all have the following properties in common: Input to the regulatory region of one gene (represented by the small black arrow) drives a positive feedback loop that turns on one or more genes in the small module, and stabilizes the new state of those downstream genes, which in turn regulate other genes that will change the state of the cell. Once these circuits are triggered, they switch inexorably to a new state and don't return to their initial state. The small boxes in the figure indicate in simplified binary form (1 is "on," 0 is "off") the initial state of the circuit (upper line) and the final state of the circuit (lower line).

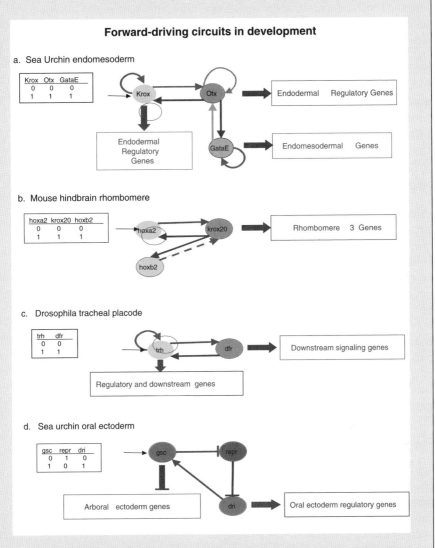

SOURCE: Figure courtesy of David Galas, based on information contained in Davidson (2003).

> If you have an apple and I have an apple and we exchange apples then you and I will still each have one apple. But if you have an idea and I have an idea and we exchange these ideas, each of us will have two ideas.

Clearly, there will be situations in which sharing information devalues that information—for example, if a person shares the location of a limited food source with others, the information sharing is likely to reduce the amount of food that the person gets from that location. But devaluing information by sharing is not an inherent property of the information itself; rather, it is a consequence of the situation. In other cases, sharing information can carry no such costs. If a person warns others about tomorrow's severe weather, the warning does not impinge upon the person's own ability to take appropriate precautions.

In some cases, sharing information may even *increase* the value of that information. For example, Marzluff et al. (1996) and Wright et al. (1998) provided compelling evidence that the communal roosts of common ravens (*Corvus corax*) serve as "information centers" in which individuals share information about the location of food sources. In this case, there are direct benefits to sharing information: The members of a communal roost cannot feed unless they arrive at the food source in large enough numbers to displace the local territory holders. Thus, knowledge of the location of a food source is useless unless shared. Moreover, the costs of sharing the information are small or nonexistent. These food resources, typically large carcasses, are often so big that a group of ravens cannot consume one entirely before the resource is lost to snowfall, mammalian scavengers, or other causes.

INFORMATION AND EVOLUTION

The evolutionary process itself can be conceptualized as a process of information acquisition. The sorts of information that are represented in the genome and the ways in which this information is extracted from the genome by the living organism were discussed earlier. But how did this information initially get into the genome? The answer is that information accumulates in the genome as a result of the process of evolution by natural selection. Mutation in its many forms provides a wide range of variation, but on its own, mutation does not necessarily add further information with respect to the environment (i.e., it does not increase Shannon's mutual information between genome and environment). For example a "silent" mutation does not immediately change an organism's phenotype. The additional information comes in as a result of the sorting process of natural selection. Selection preserves those genotypes that operate more effectively in the environment and discards those genotypes that are less effective.

One can quantify this relationship; Haldane (1957) and Kimura (1961) established that information can accumulate in a sexual population at a rate no higher than $-\log(1-s)$ bits per generation, where s is the selective load (basically, the fraction of the population lost to selection). Recent analyses of evolution in fluctuating environments (Bergstrom and Lachmann, 2004; Kussell and Leibler, 2005) further draw out the relation between theoretic measures of genomic information and the concept of Darwinian fitness. These analyses hint that the two different ways of measuring information—the Shannon framework and the decision theory framework—could be closely related under special circumstances. Kelly (1956) characterized one such set of circumstances; he established a relationship between the value of side information to a gambler and the entropy rate of the process being wagered upon. Evolution by natural selection appears to provide another example. However, further work is needed to approach a thorough understanding of these relations.

CONCLUSIONS

An attempt to characterize living systems by citing just two essential properties would probably include, first, that they are thermodynamically far from equilibrium, and second, that they store, accumulate, and transmit large amounts of information. While there is still a struggle to shape the concepts in ways that are rigorous and useful for biology, biologists can recognize that information is indeed a valuable way to describe many life processes. There are many nonbiological disciplines, including mathematics, computer science, and statistics, that have problems similar to some of those that biologists grapple with. The problem of understanding biological information and developing fruitful theoretical ideas and useful tools will likely be aided by this rich vein of ideas and methods.

8

What Determines How Organisms Behave in Their Worlds?

Organisms as diverse as bacteria and humans shape their behaviors in response to particular environmental variables. Understanding life requires determining the rules that govern how organisms behave in their worlds, how they sense their environments, and how they use this information to change their behavior. Organisms do not passively wait for information from their environments; rather, their physiology is internally generated, by genetically determined rules, and input from the environment is used to alter the behavior of the organism. Much behavior is generated to actively explore the environment in search of specific sensory signals. For example, bacteria sense changes in the concentrations of chemicals in their environment and use these to govern their movements. The integration of sensory information into a form that can be processed by the organism, the nature of the processing machinery, the influence of the internal states of the organism, the influence of the experience on the future states of the organism, memory mechanisms, and many other issues have direct relevance to many different biological regimes, scales, and kinds of organisms. There is a remarkable potential for finding commonalities amid the diversity addressed by this question.

Living organisms have an extraordinarily diverse set of tools for sensing the environment. Across the entire living world, the kinds of external cues that organisms can sense are extremely varied, ranging in intensity or power across the spectra of light and sound and across many orders of magnitude. Organisms are able to differentiate among thousands of chemicals by taste and smell and are sensitive to minute changes in temperature, pH, air speed, surface texture, and chemical concentrations. In short, there seem to be few physical parameters that are not sensed by some living organism. Elephants hear sounds of much longer wavelength than humans and have specialized cells in their feet to sense seismic vibration. Their vibrational sense is specific enough to distinguish vibrations with different meanings, and the elephants react differently depending on whether the signal is coming from a familiar elephant or a stranger (O'Connell-Rodwell et al., in press). Plants acquire information about day length and temperature, and in temperate climates they use this to time their budding and flowering. Organisms from bacteria to humans use light and other cues to entrain circadian, or daily, rhythms so that these internally generated rhythms are synchronized to the fluctuations in their environment (Nakajima et al., 2005).

The ability to detect different environmental signals and adjust behavior accordingly is evolutionarily ancient. Bacteria produce many small molecule signals that are used for communication both within and among species (Bassler and Losick, 2006). The realization that single-celled organisms—often perceived as "primitive" and "simple"—live in complex mixed-species communities and use a variety of chemical signals to detect community density and composition has stimulated a reexamination of our theories regarding the basic parameters of bacterial life. Theoretical approaches to understanding how organisms sense and respond are likely to be profitably employed across biological scales.

The ability to receive and process external cues also plays a critical role in multicellular development. For a single fertilized egg to develop into a highly differentiated and organized multicellular organism, individual cells must receive cues about their location and future role in the organism to migrate to the right place and differentiate into the appropriate specialized cell. As each of the organism's cells contain the same genetic material, the interplay between external cues, the triggering of various genetic pathways, and the subsequent modification of the cell to be able to respond to different external cues (e.g., by the expression of receptors or ion channels) results in an intricate and tightly regulated cascade that reliably produces a functional multicelled organism.

Humans and other "higher animals" have complex brains that allow them to acquire information from the environment, compare this information with both memories of prior experience and internal models of the world, and then respond, often appropriately but in the case of humans too often inappropriately for the world of 2007. For example, many humans find themselves overeating, gambling, or otherwise engaging in counterproductive behaviors, as mechanisms that have evolved over millions of years for survival in simpler social and environmental circumstances are evoked by the stimuli and circumstances of today's world. But humans are not the only organisms that may find themselves responding inappropriately in unusual environmental conditions. In the early winter of 2006-2007, unseasonably warm temperatures in the northeastern United States led to many plants prematurely budding and flowering, only to be damaged when the normal cold weather finally came.

At a still "higher" level of organization, organisms may self-associate into colonies, tribes, and other groupings in which individuals take on specialized roles to assess the environment, make decisions, and take action. For example, social insects, such as ants and bees, have fascinating patterns of divisions of labor that require not only that the individual animal acquire information for its own behavior but also that this information be communicated and shared in a larger population.

BEHAVING IN THE WORLD: FIVE MULTISCALE QUESTIONS

All living entities, be they cells within an organism, plants, bacteria, leeches, or humans, integrate information from their external and internal environments and respond, in most cases, appropriately. In this section, examples from a variety of biological organisms are used to address the following questions:

1.   How are external stimuli transduced into some kind of code that can be acted upon by the organism? How do these codes vary with the intensity, duration, and timing of the stimulus?

2.   How does the internal state of the organism influence the interpretation of sensory codes?

3.   How is past experience represented in the internal state of the organism? In other words, what kinds of memory mechanisms exist, how are they created, how long do they last, and how are they read out?

4.   How are representations of the external world combined with internally generated signals to allow the organism to integrate past and present stimuli to make decisions about relevant actions? How does memory influence decisionmaking?

5.   How are decisions used to implement specific actions? When are actions internally generated, and when are they triggered by specific events in the internal or external environment?

SPECIFIC PROBLEMS, OPPORTUNITIES, AND CHALLENGES

In each of the sections below, a few examples are given of how each question applies across biological scales, to illustrate how the same or very similar problems arise at levels ranging from the individual microbe or cell to a complex multicelled organism. Integrating knowledge across these scales will require much further study and further development of both theory and technology.

While these essential questions are easily recognized as central to the field of neuroscience, they are actually common to all biological systems, including plants, the immune system, bacteria, and fungi. Indeed "memory" is a fundamental feature of the immune system, and every living cell responds to environmental signals and encodes this information for eventual response. That said, more complex organisms require highly specialized nervous systems to do more and more complex computations to make more sophisticated responses to their environments.

1. How are external stimuli transduced into some kind of "code" that can be read by the organism? How does this code vary with the intensity, duration, and timing of the stimulus?

Cellular behavior is altered by signals from the cell's surroundings. Different types of information can pass via mechanisms involving ion channels, gap junctions, or the initiation of intracellular signals resulting from binding or clustering of transmembrane proteins. Cells can also alter their environment, for example, by secreting proteins that are assembled into an extracellular matrix. That matrix can, in turn, influence intracellular organization such as the arrangement of the cytoskeleton. Cells can also communicate with their neighbors. For example, in plants, plasmodesmata allow cytosplasmic connections through cell walls. The cell's reaction to external signals through these various mechanisms may differ depending on when the signal arrives (e.g., during darkness or light), how long the signal lasts (e.g., how long the temperature remains below freezing), and how strong the signal is (e.g., how many receptors are bound at the same time).

In "higher" animals, many stimuli are detected initially by only a small minority of cells, and specialized endocrine and neuronal systems are used to coordinate the organism's response to the stimulus. Animals have highly diverse and specialized sensory structures that allow them to turn a variety of such stimuli as light, sound, heat, body position, pH, and CO_2 into neuronal signals that eventually are integrated with internally generated signals to result in behavior. Perhaps surprisingly, it has only been quite recently that the receptors for many of these environmental stimuli have even been identified. For example, a variety of transient receptor potential (TRP) channels that respond to heat or painful stimuli have only been recently identified (Julius and Basbaum, 2001; Ramsey et al., 2006), and while many signal transduction pathways are well characterized, those activated by many *sensory* modalities remain mysterious. A beautiful example of the interaction between physics and biology can be seen in elegant work elucidating the fundamental mechanism by which sound results in hair cell deformation and changes in membrane conductances (Hudspeth, 2001; Chan and Hudspeth, 2005a, b; Hudspeth, 2005; Keen and Hudspeth,

2006; Lopez-Schier and Hudspeth, 2006; Kozlov et al., 2007). This work requires complex calculations of the forces accompanying the movements of molecules, as well as the technology to work with small and fragile biological preparations.

As suggested in Chapter 7, information theory has been usefully applied in recent work on sensory processing to study how the sensory signal is encoded in spike trains (Fairhall et al., 2001; Lewen et al., 2001; Adelman et al., 2003; Thomson and Kristan, 2005). Most early studies of sensory processes used extremely simple, well-defined, and artificial stimuli, such as spots of light or pure tones. Because animals do not spend their lives experiencing pure, well-defined stimuli, investigators are beginning to ask how sensory systems respond to natural stimuli, which change in complex and unpredictable fashions. This is much more difficult than working with simpler stimuli, as it requires characterizing the properties of the stimuli and understanding how they are captured in a spike train or a series of spike trains. This is a very new area of investigation in which a theoretical approach might be helpful. Indeed, recent years have seen the impact of Bayesian statistics on problems of neural coding, illustrating the importance for biology of quantitatively trained investigators of all kinds.

Responses to stimuli, whether artificial or natural, always show a degree of trial-to-trial variability in the responses of single neurons or groups of neurons to repeated presentations (Billimoria et al., 2006). Is this noise or is this an important feature of how the sensory world is represented? Indeed, working out the means by which different biological systems filter or sort different stimuli is another challenge. There are a number of important theoretical problems associated with understanding how noise in spike trains is dealt with by nervous systems.

The field of sensory neurophysiology provides fascinating examples of the diversity of mechanisms that animals have evolved to sense their worlds (Box 8-1). For example, electric fish live in murky waters, where vision is essentially impossible, and use electrical discharges to locate their prey and each other (Zakon and Dunlap, 1999; Zakon et al., 2002; Bass and Zakon, 2005). Some bats capture prey with the help of wideband biosonar sounds that they emit and then use to calculate the distance to objects from the delay of echoes (Simmons et al., 2004).

Sometimes the sensory response system involves more than one species. For example, the bobtail squid houses bioluminescent bacteria in specialized organs where they provide camouflage in different light conditions (Koropatnick et al., 2004). In fact, a large body of evidence is accumulating that most animals rely on a closely associated microbial community for a variety of functions, some of them sensory, such as alerting the organisms to the presence of pathogens, and detecting and degrading toxins. And, of course, there are many situations in nature whereby one species' reaction to environmental cues is interpreted and acted upon by other species (see Box 8-2). Whether these arrangements are cooperative or an example of animals expanding their own sensory repertoire to include the interpretation of other species' signals is an interesting theoretical question.

2. How does the internal state of the organism influence interpretation of the sensory "code"?

Internal conditions also influence how cells react to stimuli. Internal conditions can vary depending on the stage of development of the organism or the types of contacts with neighboring cells. For example, signals from the Notch family of proteins cause distinct changes at two different stages in the development of the nematode *C. elegans*. During embryonic development, Notch signals lead to mesoderm induction, whereas during postembryonic development they lead to germ cell mitosis (Austin and Kimble, 1987). In Box 8-3 an example is given of plant seeds that express a sensitive light receptor when they are deprived of light; when the receptor is activated by even a brief light exposure, the seed begins to germinate. Across biological scales, the response to environmental cues can differ depending on the state of the cell or organism.

Whether people experience a stimulus as painful depends to a large degree on prior history with the stimulus, expectation of its duration, and whether it is viewed as innocuous or as a portend of dire consequences. This is just one example that illustrates that internally generated neuronal activity plays important roles in shaping the processing and interpretation of sensory stimuli. Indeed, in a remarkable new study, the estimate is that internally generated activity is much more significant than the external stimulus in shaping the receptive fields of neurons responding to natural images (Fiser et al., 2004). This study is part of a newly developing area of research in which methods such as Bayesian inference are being used to understand visual processing and decision making (Ma et al., 2006). This is a very new area in neuroscience and one in which the use of theoretical methods for understanding the nervous system is needed.

Circadian rhythms are found in organisms from bacteria to humans. A great deal is now known about the sequence of molecular events that gives rise to circadian rhythmicity (Allada et al., 2001; Hardin, 2005). Circadian rhythms, by definition, are internally generated, but are normally reset and entrained by light and other environmental cues (Stoleru et al., 2004). In all organisms, there are mechanisms by which information about light and other environmental cues is used to change the phase of the internally generated molecular clock (Gehring and Rosbash, 2003). Moreover, there are mechanisms by which the output of the molecular clock can be read out to trigger changes in behavior (Stoleru et al., 2005). The circadian system demonstrates that the state of intracellular signal transcriptional

Box 8-1
Animals as Engineers:
Specialized Senses for Communication and Predation

The Jamming Avoidance Response in Weakly Electric Fish

If you are a fish living in muddy and murky waters, how do you locate your prey and your mate? The weakly electric gymnotoid fish such as Eigenmannia produce a very precise sinusoidal electric discharge (Heiligenberg, 1991). The fish use this discharge to navigate and locate their prey, as they sense reflections of the discharge by objects in their environment (Lewis and Maler, 2001). But remarkably, to avoid confounds produced by electric discharges of other fish, when two animals come into range of each other, if their discharges are close in frequency, which would effectively jam their signals, each fish alters the period of its discharge, so that the two animals now are operating at frequencies enough different so that they don't interfere with each other. This has been termed the "jamming avoidance response" and is biological bandwidth sharing that allows multiple animals to navigate simultaneously.

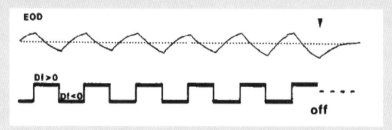

The jamming avoidance response of electric fish. Top trace, the frequency of the electric organ discharge, in response to a jamming signal turned on and off. The dotted line is the control signal in the absence of the perturbing influence. Modified from Metzner (1993).

and translational mechanisms can directly alter an organism's response to a stimulus. This is a counterexample to the common perception that organisms passively wait for input from the environment, rather than that behavior reflects an interaction between internal and external factors. There are many other examples of "counting" or "timing" mechanisms whereby cells

Bat Echolocation

Like electric fish, bats hunt their prey under very difficult conditions. Bats hunt at night, in the absence of much light. Moreover, bats hunt rapidly moving objects, most notably moths and other flying insects that are moving in three dimensions in highly unpredictable trajectories. How then do bats successfully compute the appropriate trajectories to find their prey? Many species of bats produce and sense sonar, so again like the electric fish example above, the animal produces a signal and uses the disturbed reflection of the signal to find objects. Even more remarkably, bats compensate for the Doppler shift produced by the moving insect by changing the frequency of their own calls, to maintain the frequency of the reflected signal in the range at which the bat's auditory system is optimally functional (Smotherman and Metzner, 2003). This "Doppler-shift compensation" behavior significantly enhances bats' echolocation performance in their natural habitat.

SOURCE: Auditory Adaptations for Prey Capture in Echolocating Bats, Vol. 70 by G. Neuweiler. Copyright 1990. Reproduced with permission of American Physiological Society via Copyright Clearance Center.

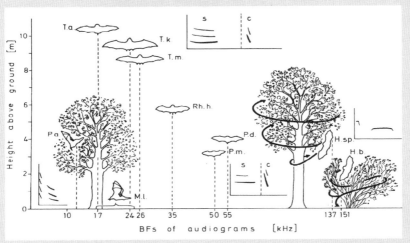

Correlation between best foraging habitat and frequency of the bat's calls. Modified from Neuweiler (1990).

maintain a record of past events (see Box 8-4), and theoretical approaches may identify common features of these mechanisms.

3. How is past experience represented in the internal state of the organism? In other words, what kinds of memory mechanisms exist, how are they created, how long do they last, and how are they "read out"?

Box 8-2
Taking Cues From Other Species

Photo taken by Rayko Halitschke.
A wild tobacco plant (front) growing next to sagebrush (background) in the Great Basin Desert in Utah.

When damaged by insects, sagebrush releases a suite of chemicals that alert neighboring plants to the presence of the insects. The signal are not only picked up by other sagebrush, though. Wild tobacco plants growing within range of the chemical signals will also stimulate the defense mechanisms they use to repel plant-eating insects like caterpillars. SOURCE: Baldwin et al. (2006).

In the broadest of terms, "memory" can be defined as a lasting trace of the prior history of the system's experience. Memory can be seen in the immune system as evidence of prior exposure to antigen. In any cell, memory can be described as the state of all of the signal transduction and gene expression networks in a cell. In both the immune system and the nervous system, memory can be quite specific to the details of the experience that triggered the memory, and in both of these systems the "memory" can be quite long-lasting and may persist much longer than the lifetime of any of

Top row: The hornbill (left) and the Diana monkey (right). Bottom row: Leopard (left) and crowned eagle (right).
Photo credits: Ben Wang, David Jenny, and the Peregrine Fund.

Diana monkeys have two different, but very similar, predator warning calls for leopards and eagles. Hornbills, which are threatened by eagles, but not by leopards, respond only to the eagle warning cry. SOURCE: H. J. Rainey, K. Zuberbuhler, and P. J. Slater. Hornbills Can Distinguish Between Primate Alarm Calls. Proceedings of the Royal Society B Biological Sciences. 2004. 271:755-759.

the molecules by which the memory is expressed. There is promise in theoretical approaches to studying the origin and maintenance of mechanisms by which organisms store and access information about the past.

Understanding the storage of memory has been and continues to be one of the most intensively studied problems in neuroscience. Although the molecular and cellular mechanisms underlying stable changes in synaptic transmissions have progressed dramatically in the past decade (Kandel, 2001), even at the subcellular level, much remains to be understood that

---

Box 8-3
Germination of Weeds After Plowing

When agricultural fields are plowed, many weeds germinate. Oddly, some-times more weeds germinate if the fields are plowed during the day than if they are plowed at night. The reason appears to be that some seeds that have experienced a period of light deprivation up-regulate an extremely sensitive light receptor that can detect minute flashes of light and trigger germination (Scopel et al., 1991). Thus even a very brief exposure to sunlight during plowing primes these seeds for germination—an exposure that does not occur if fields are plowed at night.

This mechanism may allow the plant to perceive when the soil has been disturbed and therefore favorable for growth. The phenomenon does not appear to be universal, with the effect of nighttime plowing varying according to the type of weed, the seeds' dormancy status, seasonal variation in light, soil moisture levels, and method of plowing (Juroszek and Gerhards, 2004). Its existence, however, is an example of how the internal state of a cell (whether or not the sensitive light receptor is expressed) can affect how a signal is acted upon.

Box 8-4
How Do Cicadas Know That 17 Years Have Passed?

Residents of the eastern United States are familiar with the onslaught of cicadas that occurs every 17 years. The reason for this organism's extreme life cycle may have its evolutionary roots in predator avoidance, but the mechanism appears to rely on using signals from the tree roots around which the cicada nymphs are developing. Each flowering cycle is detected by the cicada nymphs as an increase in sugars and proteins flowing to the roots; somehow the nymphs keep count of the number of cycles. In an experiment by Karban et al. (2000), 15-year-old nymphs were transplanted to the roots of trees that flowered twice per year. The nymphs emerged after two flowering cycles—during year 16 instead of year 17.

---

will require new computational models, of both the molecules at the synapse (Korkin et al., 2006; Lisman, 1985; Lisman and Zhabotinsky, 2001; Miller et al., 2005) and the biochemical processing in dendritic spines. The volume of dendritic spines is small and the biochemistry that occurs in these restricted spaces takes place under conditions that defy the assumptions of most experiments done in the test tube in large volumes (see Chapter 5).

Therefore, it is not obvious how to apply the measurements done with purified proteins and known concentrations of solutes to enzymes anchored in the membrane in very small spaces (Shifman et al., 2006). For this reason, some are starting to employ Monte Carlo methods to study the organization of synapses and the biochemistry that is likely to be responsible for changes at the synapse (Franks et al., 2002; Coggan et al., 2005; Sosinsky et al., 2005).

Understanding how memories are stored in neural networks is a topic that has and continues to attract a good deal of theoretical study. Starting with the early work by Hopfield and colleagues (Hopfield, 1982, 1984, 1987; Hopfield and Tank, 1986; Tank and Hopfield, 1987), many physicists have been attracted to neuroscience by the problem of the storage of memory in artificial neural networks (Abbott and Arian, 1987). Today, models of how the brain stores memory incorporate many more recently discovered biological details in order to understand memory storage in real biological networks. That said, there is a tremendous and continuing need for tandem experimental and theoretical studies of memory.

Biological systems must be able to access and act on stored information (working memory) and to integrate information of different kinds across various time intervals. A number of recent studies on working memory have triggered a body of theoretical and experimental work on biological integrators (Seung et al., 2000). In this context, an integrator (like a capacitor in an electronic circuit that stores charge) is a mechanism that "builds up" over time and maintains information about the history of some event before it is reset. This work was stimulated by experiments in which recordings from monkeys doing working memory tasks showed sustained discharges (Fuster and Alexander, 1971). But there is a series of questions about the mechanisms used in long-term biological integrators that are relevant to a number of different biological systems.

Many biological integrators routinely handle signals that persist for milliseconds, seconds, minutes, or even hours. These are time constants that are relatively easy to understand within the context of what is known about the storage of information in changes in membrane potential or buildup and decay of intracellular metabolites. Nonetheless, there are biological integrators that work over much longer times. For example, there is ample evidence in animals that the "sleep integrator" keeps track of how much sleep the animal gets over multiple days and that animals oversleep for several days to make up for the sleep debt incurred over days and weeks. Even more puzzling are data that suggest the existence of a long-term "caloric integrator" that keeps track over weeks and months of "energy debts" incurred by caloric restriction that causes animals to overeat after periods of caloric restriction and that contributes to the difficulties that dieters have in maintaining weight loss. It is completely unclear what kinds of mechanisms would be needed to create such long-time constant integrators, and theoretical studies could help frame the question by asking whether cell-autonomous intracellular processes could, in principle, be sufficient or whether some kind of neuronal circuit would be needed. It is important to state that this kind of "memory" may be entirely different from memories that are needed for many other kinds of information storage, as these "integrator" processes can be reset at any time by the appropriate process.

4. How are representations of the external world combined with internally generated signals to allow the organism to integrate past and present stimuli to make decisions about relevant actions? How does memory influence decision making?

"Decisions" are made in all biological systems when stimuli result in some process moving past a threshold. This can be seen in the release of a hormone, an action potential, or a variety of other processes. Of course, animals can make more complex decisions, choosing or selecting among a variety of complex behaviors (Levi and Camhi, 2000). For example, should the leech swim or crawl under a set of circumstances (Esch et al., 2002; Briggman et al., 2005), or should a human walk to the post office or drive his or her car? Understanding how sets of different sensory inputs interact with the internal state of the nervous system to allow an animal to "decide" among a variety of different possible outcomes is an area of interest for study by both theorists and experimentalists (Lo and Wang, 2006; Ma et al., 2006).

5. How are "decisions" used to implement specific actions? When are actions internally generated, and when are they triggered by specific events in the internal or external environment?

Many conventional textbooks inadvertently leave the student with the notion that the nervous system is passively awaiting sensory input that will trigger a behavioral response. However, the nervous system is constantly active, and the challenge is to understand how this internally generated activity interacts with information from the environment. All regions of the brain show ongoing, internally generated spontaneous activity. Rhythmic motor patterns are generated by central pattern generating circuits (Marder and Calabrese, 1996; Marder and Bucher, 2001) that can produce rhythmic motor patterns in the absence of sensory input. The most crucial of these for life are the respiratory centers that drive breathing. Of course, sensory inputs modify the output of respiratory and other central pattern generating circuits, as is necessary for the animal to adapt its internally generated movements to its needs in the world.

Oscillations are not only important for rhythmic movements, but it is becoming clear that oscillatory processes are central to processing in virtually all brain regions (Buzsaki and Draguhn, 2004; Buhl and Buzsaki, 2005). For this reason, there is a large body of theoretical work being done, and still needed, to understand the roles of oscillatory processes in the circuits responsible for perception and cognition (Ermentrout, 1996; White et al., 1998; Sivan and Kopell, 2006). Equally needed are new studies that provide models for how voluntary movements are produced, as these will be the substrate for developing a variety of neuroprosthetic devices to enable movement.

THE IMPORTANCE OF THRESHOLDS

A characteristic feature of many of these biological processes is that they have specific thresholds, such that stimulus intensities below threshold fail to result in a response, while higher stimulus intensities produce responses. This kind of threshold is a characteristic feature of the action potential, the unit of most electrical signaling in the nervous system, but threshold behavior and amplification can also result from many signal transduction cascades, such as those triggered by hormones in individual cells.

The action potential has a number of other important features: The relationship between membrane potential, time, and channel opening is highly nonlinear. Understanding multiple nonlinear processes that work together requires the use of computational and theoretical tools, as it is often impossible to predict intuitively the outcomes of interacting nonlinear processes without computing those outcomes directly. For this reason, it is now commonplace for neuroscientists and other biologists working to understand the interactions of nonlinear processes to construct formal models, with differing degrees of realism. As introduced in Chapter 2 (p. 28), one of the most thorny practical problems facing scientists who attempt to develop formal computational models of complex biological systems is the extent to which models should attempt to capture rich biological detail or the extent to which they can legitimately "reduce" the dimensionality of the problems to be studied. This is seen in a pronounced fashion in neuronal models, where individual model neurons may be represented extremely simply as a "firing rate," or can be complex and detailed, anatomically realistic, multicompartmental models. Collaborations with mathematicians can help develop methods for dimension reduction in models. A great deal of additional theoretical work is needed to understand how much detail models must include to capture accurately the dynamics of the biological system in question.

THEORY IN NEUROSCIENCE

For a variety of historical reasons, research in neuroscience has long reflected combined experimental and theoretical approaches. For example, the Hodgkin-Huxley formulation of the action potential (Hodgkin and Huxley, 1952) remains as useful today as it was revolutionary at the time. At the same time, research in neuroscience has long been a source of insight for technological innovation. For example, the discovery that lateral inhibition in the *Limulus* eye resulted in contrast enhancement of visual images (Hartline and Ratliff, 1957, 1958) provided early information that aided in the development of computational algorithms for contrast enhancement. Likewise, many investigators interested in robotics have been inspired by the organization of insect and other invertebrate nervous systems and skeletal-muscular adaptations to locomotion (Chiel and Beer, 1997; Ayers and Witting, 2007).

Today, neuroscience remains a field in which the interaction between theory and experimental work is rich. A large number of physicists and mathematicians have been drawn into computational neuroscience over the past 20 years, motivated by the sense that the brain poses one of the biggest mysteries left to solve and by their appreciation that understanding of computations in the brain can benefit from quantitative analyses and model building (Dayan and Abbott, 2001). Recognition of the deep evolutionary roots of sensory pathways provides opportunities for collaborative theoretical and experimental research combining neuroscience, microbiology, and plant and animal physiology.

9

How Much Can We Tell About the Past—and Predict About the Future— by Studying Life on Earth Today?

Individual organisms are ephemeral, persisting over long time scales only in the form of lineages of ancestors and descendants. Most of the species that have ever lived are extinct, and yet a great deal can be learned about them by examining fossils and by studying the genomes of living descendants. From an individual organism's genome much can be learned about its parents; at the same time, only the information that passes to the offspring will be available to future generations. Thus, the collective genetic content of all the organisms on Earth represents a treasure trove of historical data and at the same time is the result of a strong winnowing process. Not everything is transmitted to future generations. In addition, extinction has removed countless genetic combinations that were adapted to the environments and communities in which they arose.

Thinking about the collective genetic reservoir in this way—as a record of the past and as the starting point for future evolution—allows one to ask some intriguing questions. Would it ever really be possible to build a "Jurassic Park," to re-create an ecosystem from the past? Could scientists ever predict what life on Earth will look like 1,000—or 1 million—years from now? Future life must evolve from the life now present, and thus, understanding the information about the past that is embodied in current organisms and ecosystems, and understanding how organisms pass that information to future generations, is a fundamental biological question.

One of the great triumphs of 20th-century biology was to work out mechanisms of the genetic inheritance system, through which information from evolutionarily successful ancestors, recorded in DNA, is passed on to subsequent generations. The recognition that DNA transmits information across generations and the development of techniques to determine DNA sequence have allowed theory and data to combine elegantly in phylogenetic analyses to describe the evolution of organisms and their component parts, including metabolic, sensory, and developmental pathways, by comparing the DNA sequences of the relevant genes. While much is known about how genetic information is gained and lost through mutation, recombination, conversion, duplication, translocation, selection, and other processes that alter genetic material in individuals and populations, much remains to be learned about the expression and regulation of genome activity that depends on inherited genetic information. The promise of classical genetic theory was the theoretical ability to predict the form and capabilities of an organism by knowing the DNA sequence of its protein-coding genes. A comprehensive understanding of the regulation and interaction of these protein products would explain the process of development, allow prediction of the connection of genotype to phenotype (including, for example, the linking of genetic variation to disease susceptibility), and serve as the palette upon which natural selection could act. Research based on this theoretical framework has indeed contributed to the success of biological research in the last few decades and enabled the development of a vibrant biotechnology industry.

At a number of levels, observational and experimental data are accumulating that suggest that this enormously successful classical framework is ripe for further expansion. This chapter discusses some of the ways in which it is becoming clear that the characteristics of offspring cannot be fully explained by the genes acquired from their parents. First, an understanding of the roles of noncoding DNA, which makes up the bulk of the genomes of many higher organisms' genomes, will be required to link genotype to phenotype (see Chapter 3). Also, a number of mechanisms other than DNA sequence—collectively designated epigenetic mechanisms—are being shown to represent additional means to pass information from cell to daughter cell, from parent to offspring. Looking beyond the inheritance mechanisms that act within species, increased exploration of the microbial world has profound implications for our understanding of how adaptive mechanisms can be inherited and shared. As introduced in Box 3-2, microbes live in complex multispecies communities where genes can be shared between distantly related organisms. Thus, genetic adaptations can spread across evolutionary lineages. Furthermore, many if not all eukaryotic organisms live in intimate association with microbial communities that provide a number of functions from nutrition to host defense, functions that are apparently coordinated over evolutionary time scales with the functions encoded by the host organism's genome. Finally, behavioral, social, and symbolic structures (such as human language) have the potential to be carried from one generation to the next. These characteristics do not exist independently of each organism's genetic and environmental context, so full understanding of inheritance will require elucidating the complex interactions among all of these potential mechanisms of transmission of characteristics across generations.

TRACES OF EVOLUTIONARY HISTORY

The genome of every organism carries many remembrances of events long past, because almost every characteristic of an organism has evolved not *ex nihil* but instead by modification of preexisting characteristics. François Jacob (1977), who shared a Nobel Prize for elucidating the mechanism of gene regulation, referred to natural selection as a "tinkerer," rather than a designer, for selection can act only on those mutations and genetic combinations that happen to arise in a population. These need not be the best possible solutions and may well be different in different populations or species, so the construction of adaptations by selection may proceed along different paths, and not to the best possible end. Historical contingency, under which the long-term trajectory of change depends on initial genetic—or environmental—conditions, undoubtedly plays a disproportionately greater role in biology than in other natural sciences.

Phylogenetic history has long been recognized as the explanation for otherwise inexplicable morphological features. For example, the human respiratory pathway, from nasal passages through the trachea to the lungs, crosses the digestive pathway in the pharynx. This is explicable not by any functional advantage—indeed, this is why humans may choke while eating—but by the evolution of lungs from the air bladder of fishes that did not originally inhale air through nostrils. The brilliant red floral display of poinsettias (*Euphorbia pucherrima*) that grace many households at Christmas-time is not a red-petaled flower, but instead a circlet of leaves, identical in structure to the normal green leaves below, that surround a cluster of small petal-less flowers. Petals were lost in the ancestor of the entire tribe Euphorbieae, and so the evolution of a red display to attract pollinating birds resulted from selection on available genetic variation, in this case in leaf pigmentation.

Historical contingency applies equally in molecular biology. The very term "genetic code" suggests that the correspondence between codons and amino acids is a consequence of early evolutionary history, not of optimality of function. Evolutionarily new functions are performed by proteins that have been modified from ancestral proteins with different functions, and in some cases by proteins that perform a new function without any modification at all. The likely role of historical contingency in this process is dramatically illustrated by the crystalline proteins that compose the eye lenses of diverse animals (Piatigorsky, 2007). In all known vertebrates, the αB-crystallin is not just related to a small heat shock protein (*shsp*); it *is shsp*, encoded by the same gene (an example of "gene sharing"). Diverse other crystallins, that differ among vertebrate groups, serve enzymatic functions in other tissues: α-enolase in turtles and lampreys; glyceraldehyde-3-phosphate dehydrogenase in geckos; lactate dehydrogenase in ducks, crocodiles, and hummingbirds; arginosuccinate lyase in reptiles and birds, and NADPH:quinone oxidoreductase in camels and guinea pigs are all crystallin lens proteins and enzymes simultaneously. So far, no convincing functional studies have explained why particular enzymes should have been recruited to serve as lens proteins in these different lineages. What is clear, though, is that any of a great many proteins can be used as crystallins and that these proteins have been recruited from preexisting metabolic pathways, rather than evolving *de novo*. Much of the study of molecular evolution consists of determining the history of origin of functionally new proteins by gene sharing, by duplication and divergence of ancestral genes, and by evolutionary assembly of chimeric proteins from ancestral modules (Graur and Li, 2000).

Historical contingency pervades all levels of biological organization. Whether or not the species structure of ecological assemblages may be said to contain a record of past information, it certainly has been profoundly affected by past evolutionary and environmental events. Blood-feeding bats inhabit the American tropics but not Africa, despite the abundance of mammalian prey; marine snakes have evolved in the Indo-Pacific but not the Atlantic Ocean; the ecology of tropical American rain forests is greatly influenced by the abundant water held high above ground in the leaf axils of epiphytic bromeliads, but no comparable plants have evolved in the Old World tropics. The shell-drilling habit (as in modern oyster drills) evolved in a Triassic gastropod lineage but was lost when this lineage became extinct in the end-Triassic mass extinction and did not originate again for another 120 million years (Fürsich and Jablonski, 1984).

Extinction has left a major imprint on contemporary life. Because the Alps prevented dispersal of many species to low latitudes during Pleistocene glaciations, plant diversity in Europe is diminished compared to other northern land areas (Latham and Ricklefs, 1993). Echoing a long history of thought in evolutionary biology, Stephen Jay Gould (1989) argued that the human species would not exist if any of a great many environmental and evolutionary events had been different in the last 500 million years.

At all levels, from the molecular to the ecological, a major research challenge is to devise theory and statistical methods that might distinguish the relative roles of historical contingency and optimal function—of chance and necessity, as Jacques Monod put it. As methods of genetic manipulation develop, heretofore impossible experiments on function will become routine (how do guinea pigs function with a turtle's lens crystallin?), and historical narrative (to which historians of human events are largely limited) may be supplemented with scientifically testable hypotheses to a greater extent than is now possible.

EPIGENETIC MEMORY WITHIN AND BETWEEN GENERATIONS

A phenomenon that complicates our ability to predict phenotype from genotype is epigenetics. Epigenetics can help explain, for example, how genetically identical organisms can have phenotypic differences. Epigenetic developmental states record cellular "memories" of the developmental state of ancestor cells (Jablonka and Lamb, 1995; Turner, 2001). Once cells differentiate, it is often important that their state of differentiation—for example, into bone, muscle, or nerve cell types—be maintained through the rest of development and adulthood. A record of the developmental state of the differentiated parent cell is associated with storing and passing through mitosis a set of epigenetic marks on the DNA and proteins within chromosomes (such as methylation marks on DNA and post-translational modifications of histone molecules). These and other epigenetic memories are "recalled" and interpreted in the offspring cell through gene expression, which is regulated by the epigenetic developmental states within the cell.

Although development is highly reproducible and usually stable and unidirectional, other epigenetic states are established in a stochastic manner and are plastic, resulting in significant variability between genetically identical individuals. Examples include phenotypic diversity displayed by monozygotic twins, stochastic epigenetic silencing of transposable elements that influence adjacent gene expression in plants and animals, position effect variegation in *Drosophila* (silencing of genes placed next to heterochromatin through translocations), and X-chromosome inactivation in mammals. The environment can modulate the establishment and maintenance of particular epigenetic states. One classic example in plants is the flowering response to cold, known as vernalization. In certain species, germination of seeds in spring requires an exposure to cold during winter. Recent work in *Arabidopsis* has revealed that the cold signal is recorded and remembered through chromatin-level control of key flowering regulatory genes (Sung and Amasino, 2005). The "memory" of winter is used to ensure growth and flowering in the spring and summer.

Epigenetic mechanisms can function during development and during the lifetime of the organism but can also be passed on to offspring, resulting in a nonclassical means of inheritance. There are several examples in animals in which a mother's diet can influence gene expression in the offspring. The best characterized is the *Agouti* coat color phenotype in the mouse. When pregnant dams are fed methyl-supplemented diets or phytoestrogens, transcription of the *Agouti* locus is suppressed and this is associated with increased DNA methylation in a regulatory sequence upstream of the gene (Cooney et al., 2002).

Epigenetic inheritance systems that can record and transmit cellular states through meiosis also exist. Some of the best characterized examples come from studies in plants, in which a number of phenomena that involve trans-sensing mechanisms and meiotic heritability of altered epigenetic states have been reported and characterized. These include paramutation, transposon, and transgene silencing (Chandler and Stam, 2004; Matzke and Birchler, 2005). Paramutation is an allele-dependent transfer of epigenetic information, which results in the heritable silencing of one allele by another. A major difference between paramutation and the heritable transmission of silencing associated with transgenes and transposons is that the newly silenced allele is capable of silencing another active allele in subsequent generations (Chandler, 2007).

Although the phenomena are best studied in plants, epidemiological studies in humans (Bennett et al., 1997) and recent work in mice (Rassoul-zadegan et al., 2006) suggest similar phenomena occur in mammals. The more frequent observation of meiotically heritable epigenetic states in plants versus animals might be a reflection of developmental differences. Plants do not set aside a germ cell lineage early in development. Instead, cells that will produce gametes differentiate late in development from somatic cells. Thus, mitotically heritable epigenetic states accumulating in plant somatic cells are often transmitted to progeny. Genomic imprinting—allele-specific gene expression depending on whether an allele is inherited from the father or mother—occurs in both plants and animals. In mice the mechanism involves establishment of methylation marks within specific DNA sequences in the parent (there are distinct maternal and paternal marks) that are retained through embryogenesis when most genome-wide methylation is reset (Wood and Oakey, 2006). In the *Arabidopsis* plant, both alleles are methylated and the maternal allele is demethylated early in embryogenesis via a specific DNA glycosylase (Choi et al., 2002).

Several potential roles for and consequences of the transfer of epigenetic information to progeny can be envisioned. As there are a number of examples in which the environment can modulate the expression state, transferring that state to progeny could be adaptive. To be adaptive, these states would have to be highly heritable, which has been shown for several examples in plants (Melquist et al., 1999; Soppe et al., 2000), including one from a natural population (Cubas et al., 1999). Although there are many examples of highly heritable states, they are potentially reversible at frequencies higher than DNA sequence changes and thus could provide mechanisms for exploring optimum states, which might be later fixed by slower genetic evolutionary processes. High rates of change, responsiveness to environmental inputs, heritability in nongenetic inheritance systems, and their modes of interaction with genetic inheritance call for the expansion of genetic theory to understand the developmental, genetic, ecological, and evolutionary dynamics of living systems in this expanded context. Allele-specific interactions such as paramutation could also contribute to generating functional homozygosity in polyploids and might have evolved from defense mechanisms targeting viruses and other invasive genomes as some mechanistic details, such as a role for RNA-mediated chromatin changes, are shared. Paramutation-like phenomena could contribute to the low penetrance and other aspects of non-Mendelian inheritance frequently observed for genes involved in complex human diseases and the segregation of quantitative characters in other organisms.

THE CHALLENGE OF HORIZONTAL
GENE TRANSFER AND SYMBIOSIS

Contemporary phylogenetic inference—inferring the genealogy of species from records stored in morphology and molecules—is built on the assumption that life is monophyletic, so that histories of particular groups and of all life are tree-like branching structures that can be traced back to a common ancestor. As introduced in Chapter 3, evidence of wholesale and continuing lateral gene transfer within and among the three major domains of life complicates phylogenetic inferences about the earliest stages of life on Earth (true bacteria, archaea, and eukarya; see Woese, 1998; Doolittle, 1999a, b; Felsenstein, 2004). The Tree of Life Web Project at the University of Arizona acknowledges the theoretical challenge, noting that "the monophyly of Archaea is uncertain, and recent evidence for ancient lateral transfers of genes indicates that a highly complex model is needed to adequately represent the phylogenetic relationships among the major lineages of Life" (http://www.tolweb.org). Some argue that new kinds of phylogenetic theory and approaches are needed (Woese, 1998). Woese draws inspiration from physics, developing an analogy between evolution with large amounts of lateral gene transfer and physical annealing. He concludes that "the universal phylogenetic tree, therefore, is not an organismal tree at its base but gradually becomes one as its peripheral branchings emerge. The universal ancestor is not a discrete entity. It is, rather, a diverse community of cells that survives and evolves as a biological unit [made cohesive by extensive lateral gene transfer]. This communal ancestor has a physical history but not a genealogical one." Whether Woese's radical conclusions are correct or not, it appears that extensions of phylogenetic theory and possibly different methods of analysis may be needed for earliest life than for later historical periods.

Bacteria live in diverse communities where communication via small molecules and genetic exchange through several mechanisms, including lateral gene transfer and bacteriophage infection, are important to community behavior and survival. Until recently, the tools to study these complex interactions have been limited, and both data and theory needed to understand the rules of bacterial community interactions are inadequate. However, the realization that communities of single-celled organisms have mechanisms for storing information about the past has immediate relevance. Antibiotic resistance provides an example of a phenotype that can emerge in one bacterial species, be maintained in the absence of the antibiotic, and re-emerge and be shared with other bacterial species under selective pressure. Further development of theories explaining how genetic exchange within and among species provides the means to access a memory bank in different bacterial communities would have numerous applications. Are all the metabolic, defensive, and communication adaptations developed in any bacteria available to all others, making the global bacterial genetic pool a repository of all surviving past adaptations? Or are some genetic pathways, species, or communities isolated or limited in their capacity to share or be shared?

It is also becoming increasingly clear that many, if not all, eukaryotes live in close association with more or less complex communities of bacteria and archaea. How these communities assemble, the degree to which their composition is inherited, and the roles they play in the fitness of their host are only beginning to be imagined, much less described. These phenomena also seem to suggest that memory can be created at a different level—that of the community of gene-exchanging units. Lateral gene transfer early in the history of life and throughout the history of the domain of true bacteria and archaea, as well as the prevalence of symbioses in eukarya, cloud the genealogical record of biochemical pathways. Even if there remain gene-trees, these phenomena of information exchange, distributed storage, and sharing complicate current methods of phylogeny reconstruction and raise the possibility that the extension of evolutionary theory will be needed to take these phenomena into account.

ARE THERE INHERITANCE SYSTEMS NOT YET DISCOVERED?

Many types of "memory" phenomena involve the interplay of organism and environmental states. Just as studies of paramutation and heritable epigenetic change are leading to a new appreciation of the complexity of inheritance and the variety of memory phenomena, attention to ways in which environmental changes induced by organisms feed back to behavior, development, selection, and inheritance of organism traits may lead to the discovery of new kinds of inheritance systems. If organism-environment interactions result in feedback loops and there are sufficient combinatorial states of both organism and environment, then there is the potential for novel forms of nongenetic inheritance. If organisms alter their environments via traits that are passed to offspring, and environments are correlated, then offspring states (behavioral preferences, morphologies, and internal states) acquired in development and learning will record information about the historical environment in a way that potentially exposes the organism-environment interaction itself to natural selection, even if the traits are not genetically determined (Oyama et al., 2001).

For example, food preferences of parents can expose offspring to a particular range of tastes, smells, and sights associated with food (Bilkó et al., 1994; Jablonka and Lamb, 2005; see also Chapter 4). The preference for certain foods records the historical information that their rearing parent(s) preferred such foods and that such foods were present in the parent's environment (and the offspring's). Historical records of this type span many levels of biological organization, from the molecular level where odiferous molecules are transmitted to offspring (through mother's milk in lactating mammals; Galef and Henderson, 1972) to parent-offspring social interactions that bring offspring into contact with food, to organism-environment interactions including exogenous effects such as the persistence and decay of odors due to weather.

In general, the coupling of parent-offspring trait correlations (heritability) with ancestor-descendant environment correlations through organisms' alteration of their own selective environments produces a transgenerational feedback loop. In cases where the relevant traits are genetically determined, the evolution of such correlated structures is called "niche construction" (Odling-Smee et al., 2003). If traits are epigenetically (co)determined, however, an expansion of current evolutionary theory will be required to fully characterize the dynamics. Inconsistencies between observation and theory might be resolved if the possibility that environmental conditions (or their interactions with organisms) can themselves be heritable is taken into consideration. For example, the explanation of sex ratios was traditionally analyzed as a problem of the evolution of sex-determining genes, but recent studies of environmental sex determination in vertebrates hint that an expanded theoretical approach that includes organism influences on environments plus feedback to both genetic and epigenetic inheritance mechanisms may be required to fully answer this unsolved problem of evolutionary theory (see Box 9-1). Certainly such a possible feedback loop should be considered when studying the coupling of microbial community metabolism and nutrient cycles in the atmostphere and ocean, since environmental changes may affect which metabolic pathways are activated, which species predominate, and which genes are laterally transferred.

Behavioral states carry information about what has been previously experienced or learned. It is not enough, however, to record information in order to serve the inheritance function. The information must be "used" and must play a role in the growth, development, or maintenance of the

Box 9-1
Temperature-Dependent Environmental Sex Determination (TSD) and the Evolution of Sex Ratios

Classical sex allocation theory predicts an equilibrium 1:1 ratio between the sexes, assuming equal energy allocation to offspring of each sex (Fisher, 1930). Past explanations of unequal allocation appealed to a variety of genetic factors (Hamilton, 1967; see also Freedberg and Wade, 2001). Historical scenarios assumed environmental sex determination (ESD) was primitive and genetic sex determination (GSD) is derived (Ohno, 1967; see Bull, 2004), but failed to predict that highly derived groups such as vertebrates might include a mix of taxa with ESD and GSD.

The discovery of ESD in vertebrates, such as temperature-dependent sex determination (TSD) in some reptiles, challenges standard population genetics theory. ESD shows how environmental inputs to development can alter evolutionary outcomes from classical expectations (Gilbert and Bolker, 2003). The cycling of information between gene regulatory states and sex ratios to behavior, environment, and hormones and back again could result in "heritability" of the behavior variations, such as that manifested in transgenerational correlations of nest-site choice and sex ratio—a form of "non-Mendelian" behavioral or cultural inheritance (Freedberg and Wade, 2001).

Given the adaptive significance of sexual reproduction and strong selection against unequal sex ratios, theory predicts that mechanisms producing equal sex ratios should be conserved and those producing deviant ratios should be absent, yet a wide variety of ESDs are known in vertebrates (Janzen and Krenz, 2004). Bull (2004) recounts how the facts were long resisted, then accepted only as exotic exceptions, and only recently acknowledged as challenges to conventional theory. The greatest promise in solving this long-standing problem in evolutionary biology lies in integrating physiology and molecular biology with developmental biology, ecology, and evolution across many levels of organization (Valenzuela, 2004). Current approaches rely on indirect comparison of genes discovered in tractable mammal and fish systems. New approaches, such as RNA interference studies of reptilian embryos, might be needed (Bull, 2004). Progress could depend on new theory as well, as suggested by new models of sex-ratio evolution in reptiles based on alternative inheritance mechanisms, such as cultural inheritance of nest sites (Freedberg and Wade, 2001).

Gilbert (2005) identifies several pathways for environmental signal transduction into genomic regulatory responses. The endocrine system is known to be such a transducer, for example, through temperature-sensitive expression of steroidogenic factor, Sf1 (homologous to Fushi tarazu Factor 1 in *Drosophila*), in TSD turtle species. Sf1 is temperature insensitive in a GSD turtle species (Valenzuela et al., 2006).

Since maternal behavior, which is also affected by hormones, is a factor in nest choice and egg laying in turtles (Bull et al., 1988), behavior could complete an inheritance cycle linking genes, hormones, behavior, and ecology (see figure below). Gene regulatory networks could affect (1) hormone-conditioned maternal behavior, which can control (2) exposure of eggs to environmental temperature in the nest, inducing (3) steroidogenic factors to regulate genes, so as to produce

(4) a particular sex ratio in offspring. In producing female offspring, a given temperature would also set particular behavioral preferences that could in turn lead to more (or less) female offspring in the next generation.

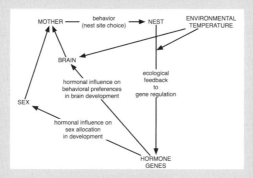

It is known that the brain plays a role in the sex determination pathway of some TSD reptiles, as the locus of transduction of temperature into hormonal signal. Aromatase, which converts testosterone to estradiol, is differentially expressed in the brain of the red-eared slider turtle, *Trachemys scripta elegans*, in a pattern that explains sex determination data in laboratory studies (Willingham et al., 2000). Thermal factors have been shown to act very early in development—for example, Sf1 acts *before* the temperature-sensitive period in *Chrysemys picta*— and that might explain why traditional genetic theory failed to predict field observations that temperature *prior* to the sensitive period can alter sex ratio (Valenzuela et al., 2006). If temperatures experienced early in development depend on maternal behavior, then behavior together with environment would close the loop in the determination of offspring sex by hormone transduction of environmental signals regulating gene expression. This possible behavioral inheritance system is not only of academic interest because TSD turtle species could be indicators of global warming. Rising average temperatures may lead to female-biased sex ratios in these species and eventual extinction (Janzen, 1994).

Although speculative, this kind of scenario points to the need to consider whole-organism behavior as well as developmental gene regulation if novel regulatory inheritance systems comparable to the genetic inheritance system are to be discovered. Behavior might be the missing link in the formation of inheritance cycles from causal chains leading from environment to hormones to gene regulation to phenotype. If so, then there may be many more types of behavioral inheritance systems than just those based on sophisticated forms of social learning. Just as developmental biology has added considerably in recent years to evolutionary biology, some argue that evo-devo (evolutionary developmental biology) needs supplementation to consider ecological aspects of development, or "eco-devo-evo" (Gilbert and Bolker, 2003). If behavioral feedback can create inheritance systems as suggested in the TSD scenario, then theorists should explore "*etho*-eco-devo-evo."

SOURCE: Courtesy of James R. Griesemer, copyright 2007.

offspring that inherit the information (along with its carriers), and thereby expose inheritance system variants of a population to evolutionary processes such as natural selection and drift.

If these nongenetic adaptations, epigenetic, behavioral, or symbolic variants are to be considered true inheritance systems, part of organisms' evolutionary legacies, they must contribute to fitness differences. Major challenges to extending investigations of nongenetic inheritance to an evolutionary context include development of new experimental tools and methods to distinguish genetic from nongenetic variation, methods of measuring fitness costs and benefits, and theory development to predict and explain evolutionary dynamics when more than one inheritance system is operating. Dual inheritance theories designed to handle cultural inheritance (e.g., Boyd and Richerson, 1985) only begin to scratch the surface of the types of inheritance systems and transmission rules involved.

CONCLUSION

Extending the concept of inheritance to include biotic and social relations implied by epigenetic mechanisms, social learning, symbolic communication through language, and interactions with environments raises questions about whether there might be a general theory of transgenerational "memory" for living systems. That is, is there a theory of biological conditions and mechanisms that record system states with the potential for closed information loops and an ontogeny and evolution of information? Just as growing awareness that genomes are dynamic complicates the simple concept that inheritance flows from genes to phenotypes and back to genes in the next generation, discoveries of other information loops (from behavior to environmental modification and back to behavior in niche construction and from symbols to social change and back to symbols in cultural evolution) present theoretical challenges as formidable as those faced by Mendel. What sorts of regular structures and mechanisms in behavior might there be to suggest transmission "rules"? What social mechanisms govern the production, manipulation, and propagation of symbols that can be captured in theories of cultural evolution? It took half of the 19th century to move from the most elementary understanding of the hereditary consequences of cross-breeding to Mendel's theory and most of the 20th century to link the implications of Mendel's theory to an understanding of the molecular mechanisms of the genetic system of inheritance in population and evolutionary processes. Scientists are only beginning to explore the mechanisms and theoretical implications of other inheritance systems and how these might interact in an expanded evolutionary dynamic.

10

Education: Learning to Think About the Elephant

Life is rich and complex. Biological study requires an interaction of theories, experiments, observations, facts, technologies, and other components. Being able to predict the behavior, fate, or ecological impact of even one organism, like the elephant introduced at the beginning of this report, requires detailed information about many of its own components, its surroundings and history. This report suggests that we can better understand the elephant by asking cross-cutting questions than by keeping our eyes closed and grasping at one part or the other of the large and complex animal. How can we learn to embrace cross-cutting questions and to increase the chances of transformative innovations? How can we learn to be self-reflective about the interplay of the many factors that make up scientific practice? How can we promote the best possible education and learning at all levels?

This committee realized that it was not our charge and further that we do not have the expertise to offer specific suggestions about content, pedagogy, textbook and other teaching materials, or learning outcomes. We do, nonetheless, point to some principles and preferred practices. Considerable research has been done on many aspects of science education, and there is wide agreement that current science education is not optimal. Too often, textbooks and standardized tests emphasize memorization of more and more facts in order to acquire content. Adding memorization of more and more theories and concepts to that mix would not help. Such an approach misses what is exciting about science and about biology and the rich diversity of the subject matter. It would be as if we were asking students to learn about the elephant by sitting in the classroom and memorizing first about atoms, then genes, cells, organ systems, and only eventually—after students are completely bored—to close their eyes and be allowed to touch the elephant.

We know from the best of education literature that science teaching works best when science is taught as science is done. Many excellent teachers show students how science is a way of knowing about the world, that it is interactive, dynamic, and exploratory, and that it draws on a mix of observations, experiments, facts, hypotheses, technologies, and theories. Everyone learns best when starting from something known and then building up facts, skills, and theoretical interpretations to arrive at better, richer, and more complex facts, skills, and theoretical interpretations.

Perhaps one reason biology education focuses on facts and observations is that being self-reflective about theory is harder. Another reason may be strategic, especially when discussing evolutionary theory, because of the need to avoid suggesting that evolution is "merely" a speculation, as many people interpret the term. Yet, as this report makes clear, theory is not mere speculation but a central and necessary part of science. It is important that biologists consciously and carefully embrace theory as essential and work to promote understanding of its central role.

One of the core theoretical foundations of all of biology is evolution, which is a theory in the sense that it is an interpretation that provides an explanation of a vast diversity of established fact. In another sense, evolution can be considered a fact since it is well established that the vastly diverse living organisms are related through common descent. The theory of evolution is so fundamental to understanding biology that no science education can be considered adequate unless students take away an appreciation of how evolution has shaped life on earth and how it acts as an organizing concept for biology.

Students at any stage come to science with experiences and background knowledge that consist of a mixture of facts and theory. They have expectations based on that experience, and they interpret their experiences in certain ways. Asking new questions allows them to recognize new facts or to discover new questions that shape new expectations and theories. In many areas of biology, mathematical and computational models are increasingly important and biology students need to be trained to go beyond arguing by assertion, to use the disciplined logic essential for the implementation of formal models to determine the adequacy of their knowledge, and to generate new hypotheses. As this report has shown, science is a complex process, and education needs both to acknowledge the complexity and to teach all aspects of the science. Science education should be about learning to recognize, evaluate, and develop new theories as well as about how to test hypotheses through well-controlled experimentation, to employ appropriate technologies, to make and record observations, and to learn and build on what are already accepted facts.

The recognition of the need for a new approach to biology education is not new. In 1875, Thomas Henry Huxley and Henry Newell Martin published the first general biology laboratory textbook, *A Course of Practical Instruction in Elementary Biology* (London: Macmillan and Co.) to provide an introduction to the principles of biological sciences through direct experience of living systems. The student was intended to dissect specimens to ask questions and discover how the parts are put together and how they work. The intention was to introduce ways of thinking in science and not just a collection of facts. In the United States, William Sedgwick and Edmund Beecher Wilson published *General Biology* in 1893 and then in 1896 introduced students directly to interactive ways of studying life.

In 1923, William Morton Wheeler reiterated that biology education must focus on living systems to bring the science alive. He pointed out that few students were choosing advanced study in biology, in large part because it was being taught badly and acknowledged that

> Any one of us who endeavors to grasp with his poor intellect, enfeebled by years of gyration in the academic mill, the stupendous and confusing accumulation of facts, not to mention the assumptions, fictions, hypotheses, theories, and dogmas that make up present-day biology, must be staggered by the difficulty of selection the most appetizing, concentrated and nourishing food for the student just entering the academic cafeteria. . . . The difficulty is greatly increased by the fact that one and all of us are highly specialized cooks, who delight in feeding the young on the dishes we ourselves like or that mother used to make and incidentally in showing our fellow cooks what delicious messes we can prepare. (Wheeler, 1923, p. 63)

To succeed with science education, as Wheeler already recognized in 1923, we need to think more carefully about how to capture students' imagination—about nature and about what science can do in studying nature.

This call for reform fits well with the exhortation to promote discovery and problem-based learning. Whether it fits with the call to "teach science as science is done" depends on which scientists we are talking about (NRC, 1996; NSF, 1998; AAAS, 2007).

We come back to our elephant. If the traditional way of studying it requires a foundation of memorized facts, then what alternatives are there? Teachers should begin from what students already know, which is experience with the diversity of life in the living world. Then education can proceed as this report does, by asking questions. Why are elephants so large? Why are baby elephants born almost two years after beginning as fertilized eggs? Why do elephants go to particular places to die? Why do elephants have the same genetic code as all other animals? What things do they eat, and does anybody eat them? If they have such big ears, why can't they fly? These are questions born of natural curiosity, and they lead students to want to know more in order to develop answers. Such questions start by embracing the complexity of biological systems and lead to discovery. The process of scientific discovery, in turn, involves observation, collection, interpretation, and theory. Life is diverse and exciting; science is diverse and exciting; science education should start from that diversity and excitement.

How science education should proceed to accomplish these goals is not a problem this committee addressed. However, it seems appropriate to support further development of pedagogical approaches, educational materials, and learning systems that recognize the complexity of the biological sciences and their interconnectedness to other systems. To teach science as it is really done, and to truly promote more effective teaching and learning at all levels from K-12 through postdoctoral training and faculty development, will require self-reflection about both how science works and how to learn to do it better.

Too often, biology courses focus on simply presenting facts and do not adequately recognize (1) the role of theory in understanding life, (2) the connections between different subdisciplines of biology, and (3) the benefits of thinking across scales. Understanding how the interplay of theory with observation, description, experimentation, technology, principles, facts, and concepts can lead to scientific advances is an important part of understanding biology. The discussion of the cross-cutting questions presented in this report illustrates how modern biological research already benefits from integration across biological disciplines and with other sciences and social sciences. The need for such integration will only increase if biology's potential contribution to answering important questions and solving practical problems is to be maximized. Yet most faculty and graduate students are trained in defined narrow disciplines and, with a few exceptions, the connections among disciplines are not explicit parts of their education. Developing a deep understanding of a particular area is an important goal of graduate education, but the most successful scientists will also understand and be able to communicate the implications of their research to those outside their area of expertise, including the general public. Biology is rich in concepts, many of which apply to multiple subdisciplines and across multiple scales (from molecule to cell to organism to population to ecosystem). Theory—in its interaction with observation, experimentation, prediction, instrumentation, and hypothesis testing—plays a key role in advancing our understanding of life and helping us to form connections between disciplines.

It is important to support faculty in efforts to alter courses and curricula in ways that are compatible with the ideas presented in this report. It would be good for undergraduate biology majors of all types and graduate students in all fields of biology to be exposed to many subdisciplines of biology and to thinking across scales of time dimension and complexity. For example, requirements to complete a major in any biology subdiscipline could include a requirement for students to take courses in other biology subdisciplines. Or question-based courses, team taught by biologists from several subdisciplines, could be developed. Courses could show how biology intersects with other sciences (chemistry, physics, mathematics, computer science, geology, engineering, and social sciences). Courses that explicitly discuss how one's theoretical and conceptual framework affects what one chooses to observe and what tools one applies guides one's experimental strategy and helps make sense of one's results and will allow students to become aware of the integral role that theory plays in the practice of biology.

11

Findings and Recommendations

Finding 1

Biological science can contribute to solving societal problems and to economic competitiveness. Basic and applied research targeted toward a particular mission is one way to accomplish this important goal. However, increased investment in the development of biology's fundamental theoretical and conceptual basis is another way to reap practical benefits from basic biological research. Theory is an integral part of all biological research, but its role is rarely explicitly recognized.

The living world presents a vast reservoir of biological solutions to many practical challenges, and biological systems can inspire innovation in many fields. The many ways that basic biological research contributes to medicine are very familiar, but basic biology can also contribute to advances in fields as diverse as food, fishery, and forest production, pest management, resource management, conservation, transportation, information processing, materials science, and engineering. Biological research breakthroughs, therefore, have the potential to contribute to the solution of many pressing problems, including global warming, pollution, loss of biodiversity, fossil fuel dependence, and emerging infectious diseases.

As the many examples in this report attest, biology is characterized by unity and diversity. There is unity because many biological processes have been preserved through evolution. There is also diversity because natural selection has led to many innovative solutions to the practical problems that living organisms have encountered over billions of years. Therefore, discoveries about a particular organism, sensory pathway, or regulatory network can have immediate applications throughout biology, and the transformative insight that provides the most direct path to a practical solution may arise in a seemingly unrelated research area. Giving explicit recognition to the role of theory in the practice of biology and increasing support for the theoretical component of biology research are ways to help make such connections and thus leverage the value of basic biological research.

The extent of life's diversity has not yet been plumbed, and many biological processes are understood only imperfectly. New tools and computational capabilities are improving biologists' ability to study complex phenomena. Tying together the results of research in the many diverse areas of biology requires a robust theoretical and conceptual framework, upon which a broad and diverse research portfolio of basic biological investigations can be based. The impact of biology on society could be enhanced if discovery and experimentation are complemented by efforts to continuously enrich biology's fundamental theoretical and conceptual basis.

Recommendation 1

Theory, as an important but underappreciated component of biology, should be given a measure of attention commensurate with that given other components of biological research (such as observation and experiment). Theoretical approaches to biological problems should be explicitly recognized as an important and integral component of funding agencies' research portfolios. Increased attention to the theoretical and conceptual components of basic biology research has the potential to leverage the results of basic biology research and should be considered as a balance to programs that focus on mission-oriented research.

Finding 2

Biologists in all subdisciplines use theory but rarely recognize the integral and multifaceted role that theory plays in their research and therefore devote little explicit attention to examining their theoretical and conceptual assumptions. Major advances in biological knowledge come about through the interplay of theoretical insights, observations, and key experimental results and by improvements in technology that make new observations, experiments, and insights possible. The fragmentation of biology into many subdisciplines means both that the mix of these components can differ dramatically from one area to another and that the development of theoretical insights that cut across subdisciplines can be difficult. It is the committee's opinion that all subdisciplines of biology would benefit from an explicit examination of the theoretical and conceptual framework that characterizes their discipline.

Recommendation 2

Biology research funding portfolios should embrace an integrated variety of approaches, including theory along with experiment, observation, and tool development. Biologists in all subdisciplines should be encouraged to examine the theoretical and conceptual framework that underlies their work and identify areas where theoretical advances would most likely lead to breakthroughs in our understanding of life. Workshops sponsored by funding agencies or scientific societies would be one way to facilitate such discussions. The theoretical and conceptual needs identified by such subdisciplinary workshops should then be integrated into the funding programs for those subdisciplines. It would also be worthwhile to sponsor interdisciplinary workshops to identify theoretical and conceptual approaches that would benefit several subdisciplines.

Finding 3

New ways of looking at the natural world often face difficulty in acceptance. Challenges to long-held theories and concepts are likely to be held to a higher standard of evidence than more conventional proposals. Proposals that break new ground can face difficulty in attracting funding, for example, those that cross traditional subdisciplinary boundaries, take a purely theoretical approach, or have the potential to destabilize a field by challenging conventional wisdom. Such proposals are likely to be perceived as "high risk" in that they are likely to fail. However, their potential for high impact warrants special attention. Successfully determining which of them deserve funding will require input from an unusually diverse group of reviewers.

Recommendation 3

Some portion of the basic research budget should be devoted to supporting proposals that are high risk and do not fall obviously into present funding frameworks. One possibility is to initiate a program specifically for such "high-risk/high-impact" proposals—whether they are purely theoretical, cross-disciplinary, or unconventional. Another is to encourage program officers to include some proportion of such proposals in their portfolios. A third is to provide unrestricted support to individuals or teams of scientists who have been identified as particularly innovative. Evaluation of these proposals should be carefully designed to ensure that reviewers with the requisite technical, disciplinary, and theoretical expertise are involved and that they are aware of the goal of supporting potentially consensus-changing research. Proposals that challenge conventional theory require not only that the originality and soundness of the theoretical approach be evaluated but also that the biological data being used are appropriate and the question being asked is significant.

Finding 4

Technological advances in arrays, high-throughput sequencing, remote sensing, miniaturization, wireless communication, high-resolution imaging, and other areas, combined with increasingly powerful computing resources and data analysis techniques, are dramatically expanding biologists' observational, experimental, and quantitative capabilities. Questions can be asked, and answered, that were well beyond our grasp only a few years ago. It is the committee's contention that an increased focus on the theoretical and conceptual basis of biology will lead to the identification of even more complex and interesting questions and will help biologists conceive of crucial experiments that cannot yet be conducted. Biologists' theoretical framework profoundly affects which tools and techniques they use in their work. All too frequently, experimental and observational horizons are unconsciously limited by the technology that is currently available. Advances in technology and computing can provide biologists with many new opportunities for experimentation and observation.

For many of the multiscale questions raised in this report, there is a strong need for teams of biologists, engineers, physicists, statisticians, and others to work together to solve cross-disciplinary problems. The interaction and collaboration of biologists with physicists, engineers, computer scientists, mathematicians, and software designers can lead to a dynamic cycle of developing new tools specifically to answer new questions, rather than limiting questions to those that can be addressed with current technology. The growing role and shortening life cycle of technology mean that biologists will have to become ever more adept in the use of new equipment and analysis techniques. Understanding the capabilities, and especially the limitations, of new instruments so that experiments are designed properly and results interpreted appropriately will be important in more and more areas of biology.

Because the potential benefits of more precise and rapid measurements of biological phenomena are so high, it will be important for biologists to be aware of both instrumentation capabilities in the physical and engineering sciences and theoretical advances in physics, chemistry, and mathematics that could be integrated into biological research. Conversely, if researchers outside biology are aware of the kinds of questions biologists are now asking, they can use their techniques, instruments, and approaches to advance biological research. Close collaboration between biologists and researchers in other fields has great promise for leveraging the value of discoveries and theoretical insights arising from basic biological research.

Recommendation 4

In order to gain the greatest possible benefit both from discoveries in the biological sciences and from new technological capabilities, biologists should look for opportunities to work with engineers, physical scientists, and others. Funding agencies should consider sponsoring interdisciplinary workshops focused on major questions or challenges (such as understanding the consequences of climate change, addressing needs for clean water, sustainable agriculture, or pollution remediation) to allow biologists, scientists from other disciplines, and engineers to learn from each other and identify collaborative opportunities. Such workshops should be designed to consider not just what is possible with current technology but also what experiments or observations could be done if technology were not an obstacle. Opportunities for biologists to learn about new instrumentation and to interact with technology developers to create new tools should be strongly supported. One possible approach would be the creation of an integrative institute focused on bioinstrumentation, where biologists could work in interdisciplinary teams to conceive of and develop new instrumentation. The National Center for Ecological Analysis and Synthesis and the National Evolutionary Synthesis Center could serve as models for the development of such an institute.

Finding 5

To get the most out of large and diverse data sets, they will need to be accessible and biologists will have to learn how to use them. While technology is making it increasingly cost-effective to collect huge volumes of data, the process of extracting meaningful conclusions from those data remains difficult, time-consuming, and expensive. Theoretical approaches show great promise for identifying patterns and testing hypotheses in large data sets. It is increasingly likely that data collected for one purpose will have relevance for other researchers. Therefore, the value of the data collected will be multiplied if the data are accessible, organized, and annotated in a standardized way. While it is somewhat new to many areas of biology, other fields that create massive data sets, like astronomy and seismology, rely on theory to guide pattern detection and to direct *in silico* experimentation and modeling. Getting the most out of the extensive biological data that can now be collected will increasingly require that biologists broadly develop those skills and collaborate with mathematicians, computer scientists, statisticians, and others. This process of building community databases is well underway in many areas of biology, genomics being a prominent example, but the specialized databases developed by one research community may be unknown or inaccessible to researchers in other fields. Significant resources are needed to maintain, curate, and interconnect biological databases.

Recommendation 5

Attention should be devoted to ensuring that biological data sets are stored and curated to be accessible to the widest possible population of researchers. In many cases, this will require standardization. Providing opportunities for biologists to learn from other disciplines that routinely carry out theoretical research on diverse data sets should also be explicitly encouraged.

References

Abbott, L. F., and Y. Arian. 1987. Storage capacity of generalized networks. *Physical Review A* 36:5091-5094.

Adelman, T. L., W. Bialek, and R. M. Olberg. 2003. The information content of receptive fields. *Neuron* 40:823-833.

Aizenberg, J., A. Tkachekno, S. Weiner, L. Addadi, and G. Hendler. 2001. Calcitic microlenses as part of the photoreceptor system in brittlestars. *Nature* 412:819-822.

Akcay, E., and J. Roughgarden. 2007. Negotiation of mutualism: Rhizobia and legumes. *Proceedings of the Royal Society B Biological Sciences* 274:25-32.

Alon, U. 2006. *An introduction to systems biology: Design principles of biological circuits.* New York: Chapman-Hill.

Allada, R., P. Emery, J. S. Takahashi, and M. Rosbash. 2001. Stopping time: The genetics of fly and mouse circadian clocks. *Annual Review of Neuroscience* 24:1091-1119.

Allen, A. P., and J. F. Gillooly. 2006. Assessing latitudinal gradients in speciation rates and biodiversity at the global scale. *Ecology Letters* 9:947-954.

Allen, A. P., J. F. Gillooly, V. M. Savage, and J. H. Brown. 2006. Kinetic effects of temperature on rates of genetic divergence and speciation. *Proceedings of the National Academy of Sciences USA* 103:9130-9135.

AAAS (American Association for the Advancement of Science). 2007. *AAAS Project 2061.* Accessed on July 22, 2007 at http://www.project2061.org/.

Ancel, L. W., and W. Fontana. 2000. Plasticity, evolvability, and modularity in RNA. *Journal of Experimental Zoology* 288:242-283.

Arabidopsis Genome Initiative. 2000. Analysis of the genome sequences of the flowering plant *Arabidopsis thaliana*. *Nature* 388:539-547.

Armour, J. A. 2006. Tandemly repeated DNA: Why should anyone care? *Mutation Research* 598:6-14.

Austin, J., and J. Kimble. 1987. GLP-1 is required in the germ line for regulation of the decision between mitosis and meiosis in *C. elegans. Cell* 51:589-599.

Avise, J. C. 2004. *Molecular markers, natural history, and evolution.* Sunderland: Sinauer Associates.

Ayers, J., and J. Witting. 2007. Biomimetic approaches to the control of underwater walking machines. *Philosophical Transactions of the Royal Society A—Mathematical Physical and Engineering Sciences* 365:273-295.

Bagemihl, B. 1999. *Biological exuberance, animal homosexuality and natural diversity.* London: Profile Books.

Baldwin, I. T., R. Halitschke, A. Paschold, C. C. von Dahl, and C. A. Preston. 2006. Volatile signaling in plant-plant interactions: "Talking trees" in genomics era. *Science* 311:812-815.

Barabási, A. L. 2002. *Linked: The new science of networks.* New York: Perseus.

Barabási, A. L., and R. Albert. 1999. Emergence of scaling in random networks. *Science* 286:509-512.

Barrett, H. C., and R. Kurzban. 2006. Modularity in cognition: Framing the debate. *Psychological Review* 113:628-647.

Barrett, P. H., P. J. Gautrey, S. Herbert, D. Kohn, and S. Smith (eds.) 1987. *Charles Darwin's notebooks, 1836-1844.* Ithaca, NY: Cornell University Press.

Bass, A. H., and H. H. Zakon. 2005. Sonic and electric fish: At the crossroads of neuroethology and behavioral neuroendocrinology. *Hormones and Behavior* 48:360-372.

Bassler, B. L., and R. Losick. 2006. Bacterially speaking. *Cell* 125:237-246.

Beadle, G. W., and E. L. Tatum. 1941. Genetic control of biochemical reactions in neurospora. *Proceedings of the National Academy of Sciences USA* 27:499-506.

Bejá, O., L. Aravind, E. V. Koonin, M. T. Suzuki, A. Hadd, L. P. Nguyen, S. Jovanovich, C. M. Gates, R. A. Feldman, J. L. Spudich, E. N. Spudich, and E. F. Delong. 2000. Bacterial rhodopsin: Evidence for a new type of phototrophy in the sea. *Science* 289:1902-1906.

Bejá, O., E. N. Spudich, J. L. Spudich, M. Leclerc, and E. F. DeLong. 2001. Proteorhodopsin phototrophy in the ocean. *Nature* 411:786-789.

Benner, S., D. Caraco, J. M. Thomson, and E. A. Gaucher. 2002. Planetary biology—paleontological, geological, and molecular histories of life. *Science* 296:864-868.

Bennett, S. T., A. J. Wilson, L. Esposito, N. Bouzekri, D. E. Undlien, F. Cucca, L. Nistico, R. Buzzetti, E. Bosi, F. Pociot, J. Nerup, A. Cambon-Thomsen, A. Pugliese, J. P. Shield, P. A. McKinney, S. C. Bain, C. Polychronakos, and J. A. Todd. 1997. Insulin VNTR allele-specific effect in type I diabetes depends on identity of untransmitted paternal allele. *Nature Genetics* 17:350-352.

Bergstrom, C., and M. Lachmann. 2004. Shannon information and biological fitness. *IEEE Information Theory Workshop 2004. IEEE(October)*:50-54.

Bernard, C. 1865. *An introduction to the study of experimental medicine.* First English translation by Henry Copley Greene. Macmillan & Co., Ltd., 1927; reprinted in 1949.

Berner, R. A. 2004. *The Phanerozoic Carbon Cycle: CO_2 and O_2.* Oxford: Oxford University Press.

Bilkó, A., A. Vilmos, and R. Hudsoni. 1994. Transmission of food preference in the rabbit: The means of information transfer. *Physiology and Behavior* 56:907-912.

Billimoria, C. P., R. A. DiCaprio, J. T. Birmingham, L. F. Abbott, and E. Marder. 2006. Neuromodulation of spike-timing precision in sensory neurons. *Journal of Neuroscience* 26:5910-5919.

Bird, C. P., B. E. Stranger, and E. T. Dermitzakis. 2006. Functional variation and evolution of non-coding DNA. *Current Opinion in Genetics and Development* 16:559-564.

Blows, M. W., and A. A. Hoffman. 2005. A reassessment of genetic limits to evolutionary change. *Ecology* 86:1371-1384.

Bolker, J. A. 2000. Modularity in development and why it matters to evo-devo. *American Zoologist* 40:770-776.

Bolouri, H., and E. H. Davidson. 2003. Transcriptional regulatory cascades in development: Initial rates, not steady state, determine network kinetics. *Proceedings of the National Academy of Sciences USA* 100:9371-9376.

Bonneau, R., D. J. Reiss, P. Shannon, M. Facciotti, L. Hood, N. B. Baliga, and V. Thorsson. 2006. The Inferelator: An algorithm for learning parsimonious regulatory networks from systems-biology data sets. *Genome Biology* 7:36.

Bonnefont, X., A. Lacampagne, A. Sanchez-Hormigo, E. Fino, A. Creff, M. Mathieu, S. Smallwood, D. Carmignac, P. Fontanaud, P. Travo, G. Alonso, N. Courtois-Coutry, S. M. Pincus, I. C. A. F. Robinson, and P. Mollard. 2005. Revealing the large-scale network organization of growth hormone-secreting cells. *Proceedings of the National Academy of Sciences USA* 102:16880-16885.

Boore, J. L. 2006. The use of genome-level characters for phylogenetic reconstruction. *Trends in Ecology and Evolution* 21:439-446.

Bowers, J. E., B. A. Chapman, J. Rong, and A. H. Paterson. 2003. Unravelling angiosperm genome evolution by phylogenetic analysis of chromosomal duplication events. *Nature* 422:433-438.

Boyd, R., and P. J. Richerson. 1985. *Culture and the evolutionary process*. Chicago, IL: University of Chicago Press.

Bradshaw, A. D. 1991. The Croonian lecture, 1991: Genostasis and the limits to evolution. *Philosophical Transactions of the Royal Society of London Series B—Biological Sciences* 333:289-305.

Brandt, 2001. Hybrid systems. In *Multiscale and multiresolution methods: Theory and applications*, edited by T. F. Chan, and R. Haimes. Berlin; New York: Springer-Verlag. P. 1-96.

Briggman, K. L., H. D. Abarbanel, and W. B. Kristan, Jr. 2005. Optical imaging of neuronal populations during decision-making. *Science* 307:896-901.

Brown, J. H., and M. V. Lomolino. 1998. *Biogeography*. Sunderland, MA: Sinauer Associates.

Buhl, D. L., and G. Buzsaki. 2005. Developmental emergence of hippocampal fast-field "ripple" oscillations in the behaving rat pups. *Neuroscience* 134:1423-1430.

Bull, J. 2004. Perspective on sex determination: Past and future. In *Temperature-dependent sex determination in vertebrates*, edited by N. Valenzuela and V. Lance. Washington, DC: Smithsonian Books. Pp. 5-8.

Bull, J., W. Gutzke, and M. Bulmer. 1988. Nest choice in a captive lizard with temperature-dependent sex determination. *Journal of Evolutionary Biology* 2:177-184.

Buss, L. 1987. *The evolution of individuality*. Princeton, NJ: Princeton University Press.

Bustamante, C. D., A. Fledel-Alon, S. Williamson, R. Nielsen, M. Todd Hubisz, S. Glanowski, D. M. Tanenbaum, T. J. White, J. J. Sninsky, R. D. Hernandez, D. Civello, M. D. Adams, M. Cargill, and A. G. Clark. 2005. Natural selection on protein-coding genes in the human genome. *Nature* 437:1153-1157.

Buzas, M. A., L. S. Collins, and S. J. Culver. 2002. Latitudinal difference in biodiversity caused by higher tropical rate of increase. *Proceedings of the National Academy of Sciences USA* 99:7841-7843.

Buzsaki, G., and A. Draguhn. 2004. Neuronal oscillations in cortical networks. *Science* 304:1926-1929.

Cartwright, N. 1983. *How the laws of physics lie*. New York: Oxford University Press.

Caughley, G. 1994. Directions in conservation biology. *Journal of Animal Ecology* 63:215-244.

Chaitin, G. 1966. On the length of programs for computing finite binary sequences. *Journal of the Association of Computing Machinery* 16:145-159.

Chan, D. K., and A. J. Hudspeth. 2005a. Mechanical responses of the organ of corti to acoustic and electrical stimulation in vitro. *Biophysical Journal* 89:4382-4395.

Chan, D. K., and A. J. Hudspeth. 2005b. Ca2+ current-driven nonlinear amplification by the mammalian cochlea in vitro. *Nature Neuroscience* 8:149-155.

Chan, S. W., I. R. Henderson, and S. E. Jacobsen. 2005. Gardening the genome: DNA methylation in *Arabidopsis thaliana*. *Nature Reviews Genetics* 6:351-360.

Chandler, V. L. 2007. Paramutation: From maize to mice. *Cell* 128:641-645.

Chandler, V. L., and M. Stam. 2004. Chromatin conversations: Mechanisms and implications of paramutation. *Nature Reviews Genetics* 5:532-544.

Chang, B. S. 2003. Ancestral gene reconstruction and synthesis of ancient rhodopsins in the laboratory. *Integrative and Comparative Biology* 43:500-507.

Chang, B. S., K. Jönsson, M. A. Kazmi, M. J. Donoghue, and T. P. Sakmar. 2002. *Molecular Biology and Evolution* 19:1483-1489.

Charlesworth, D., X. Vekemans, V. Castric, and S. Glémin. 2005. Plant self-incompatibility systems: A molecular evolutionary perspective. *New Phytologist* 168:61-69.

Cheverud, J. M., T. H. Ehrich, T. T. Vaughn, S. F. Koreishi, R. B. Linsey, and L. S. Pletscher. 2004. Pleiotropic effects on mandibular morphology. II: Differential epistasis and genetic variation in morphological integration. *Journal of Experimental Zoology Part B—Molecular and Developmental Evolution* 302:424-435.

Chiel, H. J., and R. D. Beer. 1997. The brain has a body: Adaptive behavior emerges from interactions of nervous system, body and environment. *Trends in Neurosciences* 20:553-557.

Choi, Y., M. Gehring, L. Johnson, M. Hannon, J. J. Harada, R. B. Goldberg, S. E. Jacobsen, and R. L. Fischer. 2002. DEMETER, a DNA glycosylase domain protein, is required for endosperm gene imprinting and seed viability in *Arabidopsis*. *Cell* 110:33-42.

Clements, F. E. 1936. Nature and structure of the climax. *Journal of Ecology* 24:252-284.

Coggan, J. S., T. M. Bartol, E. Esquenazi, J. R. Stiles, S. Lamont, M. E. Martone, D. K. Berg, M. H. Ellisman, and T. J. Sejnowski. 2005. Evidence for ectopic neurotransmission at a neuronal synapse. *Science* 309:446-451.

Connell, J. H. 1971. On the role of natural enemies in preventing competitive exclusion in some marine mammals and rain forest trees. In *Dynamics of populations*, edited by P. J. den Boer, and G. R. Gradwell. Wageningen, Netherlands: Center for Agricultural Publishing and Documentation. Pp. 298-312.

Cooney, C. A., A. A. Dave, and G. L. Wolff. 2002. Maternal methyl supplements in mice affect epigenetic variation and DNA methylation of offspring. *Journal of Nutrition* 132:2393S-2400S.

Corbi, N., V. Libri, A. Onori, and C. Passananti. 2004. Synthetic zinc finger peptides: Old and novel applications. *Biochemistry and Cell Biology* 82:428-436.

Crick, F. H., L. Barnett, S. Brenner, and R. J. Watts-Tobin. 1961. General nature of the genetic code for proteins. *Nature* 192:1227-1232.

Cubas, P., C. Vincent, and E. Coen. 1999. An epigenetic mutation responsible for natural variation in floral symmetry. *Nature* 401:157-161.

Darwin, C. 1859. *On the origin of species by means of natural selection, or the preservation of favoured races in the struggle for life*. London: John Murray 1st edition.

Darwin, C. 1871. *The Descent of Man and Selection in Relation to Sex*. London: John Murray.

Davidson, E. H. *Sea Urchin Genome Project*. Accessed on August 23, 2007 at http://www.its.caltech.edu/~davidson/.

Davidson, E. H., D. R. McClay, and L. Hood. 2003. Regulatory gene networks and the properties of the developmental process. *Proceedings of the National Academy of Sciences USA* 100:1475-1480.

Davies, N. B., I. R. Hartley, B. J. Hatchwell, and N. E. Langmore. 1996. Female control of copulations to maximize male help: A comparison of polygynandrous alpine accentors, *Prunella collaris*, and dunnocks, *P. modularis. Animal Behaviour* 51:27-47.

Dayan, P., and L. F. Abbott. 2001. *Theoretical neuroscience*. Cambridge, MA: The MIT Press.

De Chadarevian, S., and N. Hopwood. 2004. *Models: The third dimension of science*. Stanford, CA: Stanford University Press.

Del Sol, A., M. J. Araúzo-Bravo, D. Amoros, and R. Nussinov. 2007. Modular architecture of protein structures and allosteric communications: Potential implications for signaling proteins and regulatory linkages. *Genome Biology* 8:R92.

Dermitzakis, E. T., and A. G. Clark. 2002. Evolution of transcription factor binding sites in mammalian gene regulatory regions: Conservation and turnover. *Molecular Biology and Evolution* 19:1114-1121.

Devos, D., S. Dokudovskaya, R. Williams, F. Alber, N. Eswar, B. T. Chait, M. P. Rout, and A. Sali. 2006. Simple fold composition and modular architecture of the nuclear pore complex. *Proceedings of the National Academy of Sciences USA* 103:2172-2177.

Dewey, T. G. 1996. Algorithmic complexity of a protein. *Physical Review E* 54:R39-R41.

Dewey, T. G. 1997. Algorithmic complexity and thermodynamics of sequence-structure relationships in proteins. *Physical Review E* 56:4545-4552.

D'Hondt, S., S. Rutherford, and A. J. Spivack. 2002. Metabolic activity of subsurface life in deep-sea sediments. *Science* 295:2067-2070.

Ding, W., L. Lin, B. Chen, and J. Dai. 2006. L1 elements, processed pseudogenes and retrogenes in mammalian genomes. *IUBMB Life* 58:677-685.

Dobson, A., D. Lodge, J. Alder, G. S. Cumming, J. Keymer, J. McGlade, H. Mooney, J. A. Rusak, O. Sala, V. Wolters, D. Wall, R. Winfree, and M. A. Xenopoulos. 2006. Habitat loss, trophic collapse, and the decline of ecosystem services. *Ecology* 87:1915-1924.

Doolittle, W. F. 1999a. Phylogenetic classification and the universal tree. *Science* 284: 2124-2128.

Doolittle, W. F. 1999b. Lateral genomics. *Trends in Biochemical Sciences* 24:M5-M8.

Edwards, R. A., and F. Rohwer. 2005. Viral metagenomics. *Nature Reviews Microbiology* 3:504-510.

Ellis, J. R. 2001. Macromolecular crowding: Obvious but underappreciated. *Trends in Biochemical Sciences* 26:597-604.

Ellner, S., and P. Turchin. 1995. Chaos in a noisy world: New methods and evidence from time-series analysis. *American Naturalist* 145:343-375.

Elowitz, M. B., A. J. Levine, E. D. Siggia, and P. S. Swain. 2002. Stochastic gene expression in a single cell. *Science* 297:1183-1186.

ENCODE Project Consortium. 2007. Identification and analysis of functional elements in 1% of the human genome by the ENCODE pilot project. *Nature* 447:799-816.

Ermentrout, B. 1996. Type I membranes, phase resetting curves, and synchrony. *Neural Computation* 8:979-1001.

Esch, T., K. A. Mesce, and W. B. Kristan. 2002. Evidence for sequential decision making in the medicinal leech. *Journal of Neuroscience* 22:11045-11054.

Fairhall, A. L., G. D. Lewen, W. Bialek, and R. R. de Ruyter Van Steveninck. 2001. Efficiency and ambiguity in an adaptive neural code. *Nature* 412:787-792.

Falkowski, P. G., R. J. Scholes, E. Boyle, J. Canadell, D. Canfield, J. Elser, N. Gruber, K. Hibbard, P. Högberg, S. Linder, F. T. Mackenzie, B. Moore III, T. Pedersen, Y. Rosenthal, S. Seitzinger, V. Smetacek, and W. Steffen. 2000. The global carbon cycle: A test of our knowledge of earth as a system. *Science* 290:291-296.

Feder, M. E. 2005. Aims of undergraduate physiology education: A view from the University of Chicago. *Advances in Physiology Education* 29:3-10.

Felsenstein, J. 2004. *Inferring phylogenies*. Sunderland, MA: Sinauer Associates.

Fine, P. V., and R. H. Ree. 2006. Evidence for a time-integrated species-area effect on the latitudinal gradient in tree diversity. *American Naturalist* 168:796-804.

Fiser, J., C. Chiu, and M. Weliky. 2004. Small modulation of ongoing cortical dynamics by sensory input during natural vision. *Nature* 431:573-578.

Fischer, A. G. 1960. Latitudinal variations in organic diversity. *Evolution* 14:64-81.

Fisher, R. A. 1930. *The genetical theory of natural selection*. Oxford: Oxford University Press.

Fisher, S., E. A. Grice, R. M. Vinton, S. L. Bessling, and A. S. McCallion. 2006. Conservation of RET regulatory function from human to zebrafish without sequence similarity. *Science* 312:276-279.

Ford, M. J. 2002. Applications of selective neutrality tests to molecular ecology. *Molecular Ecology* 11:1245-1262.

Franks, K. M., T. M. Bartol, Jr., and T. J. Sejnowski. 2002. A Monte Carlo model reveals independent signaling at central glutamatergic synapses. *Biophysical Journal* 83:2333-2348.

Frausto da Silva, R. J., and J. J. R. Williams. 1996. *The natural selection of the chemical elements: The environment and life's chemistry*. New York: Oxford University Press.

Freedberg, S., and M. J. Wade. 2001. Cultural inheritance as a mechanism for population sex-ratio bias in reptiles. *Evolution* 55:1049-1055.

Fürsich, F. T., and D. Jablonski. 1984. Late Triassic naticid drillholes: Carnivorous gastropods gain a major adaptation but fail to radiate. *Science* 224:78-80.

Fuster, J. M., and G. E. Alexander. 1971. Neuron activity related to short-term memory. *Science* 173:652-654.

Futuyma, D. J., M. C. Keese, and D. J. Funk. 1995. Genetic constraints on macroevolution: The evolution of host affiliation in the leaf beetle genus *Ophraella*. *Evolution* 49:797-809.

Galef, B., and P. Henderson. 1972. Mother's milk: A determinant of the feeding preferences of weaning rat pups. *Journal of Comparative and Physiological Psychology* 78:213-219.

Galison, P. L. 1987. *Image and logic: A material culture of microphysics*. Chicago, IL: University of Chicago Press.

Gehring, W., and M. Rosbash. 2003. The coevolution of blue-light photoreception and circadian rhythms. *Journal of Molecular Evolution* 57:S286-S289.

Ghalambor, C. K., R. B. Huey, P. R. Martin, J. J. Tewksbury, and G. Wang. 2006. Are mountain passes higher in the tropics? Janzen's hypothesis revisited. *Integrative and Comparative Biology* 46:5-17.

Giere, R. N. 1988. *Explaining science: A cognitive approach*. Chicago, IL: University of Chicago Press.

Giere, R. N. 1999. Using models to represent reality. In *Model-based reasoning in scientific discovery*, edited by L. Magnani, N. J. Nersessian, and P. Thagard. New York: Kluwer/Plenum. Pp. 41-57.

Giere, R. N., J. Bickle, and R. F. Mauldin. 2006. *Understanding scientific reasoning*, 5th edition. Belmont: Thomson Wadsworth Publishers.

Gilad, Y., O. Man, S. Pääbo, and D. Lancet. 2003. Human specific loss of olfactory receptor genes. *Proceedings of the National Academy of Sciences USA* 100:3324-3327.

Gilad, Y., V. Wiebe, M. Przeworski, D. Lancet, and S. Pääbo. 2004. Loss of olfactory receptor genes coincides with the acquisition of full trichromatic vision in primates. *Public Library of Science Biology* 2:0120-0125.

Gilbert, S. 2005. Mechanisms for the environmental regulation of gene expression: Ecological aspects of animal development. *Journal of Biosciences* 30:65-74.

Gilbert, S., and J. Bolker. 2003. Ecological developmental biology: Preface to the symposium. *Evolution and Development* 5:3-8.

Gill, S. R., M. Pop, R. T. DeBoy, P. B. Eckburg, P. J. Turnbaugh, B. S. Samuel, J. I. Gordon, D. A. Relman, C. M. Fraser-Liggett, and K. E. Nelson. 2006. Metagenomic analysis of the human distal gut microbiome. *Science* 312:1355-1359.

Gillman, L. N., and S. D. Wright. 2006. The influence of productivity on the species richness of plants: A critical assessment. *Ecology* 87:1234-1243.

Gillooly, J. F., J. H. Brown, G. B. West, V. M. Savage, and E. L. Charnov. 2001. Effects of size and temperature on metabolic rate. *Science* 293:2248-2251.

Godfrey-Smith, P. 2000. On the theoretical role of "genetic coding." *Philosophy of Science* 67:26-44.

Godfrey-Smith, P. 2007. Information in biology. In *The Cambridge companion to the philosophy of biology*, edited by D. Hull and M. Ruse. Cambridge: Cambridge University Press.

Gould, S. J. 1989. *Wonderful life: The Burgess Shale and the nature of history*. New York-London: W. W. Norton.

Gould, S. J. 1994. *Hen's teeth and horse's toes*. New York: W. W. Norton. Pp. 253-262.

Graur, D., and W-H. Li. 2000. *Fundamentals of molecular evolution*. Sunderland, MA: Sinauer Associates.

Grenfell, B. T., K. Wilson, B. F. Finkenstädt, T. N. Coulson, S. Murray, S. D. Albon, J. M. Pemberton, T. H. Clutton-Brock, and M. J. Crawley. 1998. Noise and determinism in synchronized sheep dynamics. *Nature* 394:674-677.

Grewal, S. I., and S. C. Elgin. 2007. Transcription and RNA interference in the formation of heterochromatin. *Nature* 447:399-406.

Griesemer, J. R. 2000. Development, culture and the units of inheritance. *Philosophy of Science* 67(Proceedings):S348-S368.

Griesemer, J. R., and M. J. Wade. 1988. Laboratory models, causal explanation and group selection. *Biology and Philosophy* 3:67-96.

Hacking, I. 1983. *Representing and intervening: Introductory topics in the philosophy of natural science*. Cambridge: Cambridge University Press.

Haig, D. 1993. Genetic conflicts in human pregnancy. *Quarterly Review of Biology* 68:495-532.

Haldane, J. B. S. 1957. The cost of natural selection. *Journal of Genetics* 55:511-524.

Hamilton, W. D. 1967. Extraordinary sex ratios. *Science* 156:477-488.

Han, J. D., N. Bertin, T. Hao, D. S. Goldberg, G. F. Berriz, L. V. Zhang, D. Dupuy, A. J. M. Walhout, M. E. Cusick, F. P. Roth, and M. Vidal. 2004. Evidence for dynamically organized modularity in the yeast protein-protein interaction network. *Nature* 430:88-93.

Hanada, K., X. Zhang, J. O. Borevitz, W. H. Li, and S. H. Shiu. 2007. A large number of novel coding small open reading frames in the intergenic regions of the *Arabidopsis thaliana* genome are transcribed and/or under purifying selection. *Genome Research* 17:632-640.

Hansen, T. F. 2003. Is modularity necessary for evolvability? Remarks on the relationship between pleiotropy and evolvability. *Biosystems* 69:83-94.

Hardin, P. E. 2005. The circadian timekeeping system of *Drosophila*. *Current Biology* 15: R714-722.

Harrison, P. M., H. Hegyi, S. Balasubramanian, N. Luscombe, P. Bertone, T. Johnson, N. Echols, and M. Gerstein. 2002. Molecular fossils in the human genome: Identification and analysis of the pseudogenes in chromosomes 21 and 22. *Genome Research* 12:272-280.

Hartline, H. K., and F. Ratliff. 1957. Inhibitory interaction of receptor units in the eye of Limulus. *Journal of General Physiology* 40:357-376.

Hartline, H. K., and F. Ratliff. 1958. Spatial summation of inhibitory influences in the eye of Limulus, and the mutual interaction of receptor units. *Journal of General Physiology* 41:1049-1066.

Hershey, A. D., and M. Chase. 1952. Independent functions of viral protein and nucleic acid in growth of bacteriophage. *Journal of General Physiology* 1:39-56.

Heiligenberg, W. 1991. The neural basis of behavior: A neuroethological view. *Annual Review of Neuroscience* 14:247-267.

Himmelreich, R., H. Hilbert, H. Plagens, E. Pirkl, B. C. Li, and R. Herrmann. 1996. Complete sequence analysis of the genome of the bacterium *Mycoplasma pneumoniae*. *Nucleic Acids Research* 24:4420-4449.

Hinman, V. F., A. Nguyen, R. A. Cameron, and E. H. Davidson. 2003. Developmental gene regulatory network architecture across 500 MYA of echinoderm evolution. *Proceedings of the National Academy of Sciences USA* 100:13356-13361.

Hlavacek, W. S., J. R. Faeder, M. L. Blinov, R. G. Posner, M. Hucka, and W. Fontana. 2006. Rules for modeling signal-transduction systems. *Science Signal Transduction Knowledge Environment* 2006:re6.

Hodgkin, A. L., and A. F. Huxley. 1952. A quantitative description of membrane current and its application to conduction and excitation in nerve. *Journal of Physiology* 117:500-544.

Hoffmann, A. A., R. J. Hallas, J. A. Dean, and M. Schiffer. 2003. Low potential for climatic stress adaptation in a rainforest *Drosophila* species. *Science* 301:100-102.

Hoffmeister, M., and W. Martin. 2003. Interspecific evolution: Microbial symbiosis, endosymbiosis and gene transfer. *Environmental Microbiology* 5:641-649.

Holt, R. D., and R. Gomulkiewicz. 2004. Conservation implications of niche conservatism and evolution in heterogeneous environments. In *Evolutionary conservation biology*, edited by R. Ferrière, U. Dieckmann, and D. Couvet. Cambridge: Cambridge University Press. Pp. 244-264.

Hood, L., and D. J. Galas. 2003. DNA's digital code. *Nature* 421:444-448.

Hopfield, J. J. 1982. Neural networks and physical systems with emergent collective computational abilities. *Proceedings of the National Academy of Sciences USA* 79:2554-2558.

Hopfield, J. J. 1984. Neurons with graded response have collective computational properties like those of two-state neurons. *Proceedings of the National Academy of Sciences USA* 81:3088-3092.

Hopfield, J. J. 1987. Learning algorithms and probability distributions in feed-forward and feed-back networks. *Proceedings of the National Academy of Sciences USA* 84:8429-8433.

Hopfield, J. J., and D. W. Tank. 1986. Computing with neural circuits: A model. *Science* 233:625-633.

Howard, M. L., and E. H. Davidson. 2004. *cis*-regulatory control circuits in development. *Developmental Biology* 271:109-118.

Hubbell, S. P. 2001. The unified neutral theory of biodiversity and biogeography. In *Monographs in population biology*, edited by S. A. Levin and H. S. Horn. Princeton, NJ: Princeton University Press.

Hudspeth, A. J. 2001. How the ear's works work: Mechanoelectrical transduction and amplification by hair cells of the internal ear. *Harvey Lectures* 97:41-54.

Hudspeth, A. J. 2005. How the ear's works work: Mechanoelectrical transduction and amplification by hair cells. *Comptes Rendus Biologies* 328:155-162.

Humphreys, P. 2004. *Extending ourselves: Computational science, empiricism, and scientific method*. New York: Oxford University Press.

Huston, M. A. 1994. *Biological diversity: The coexistence of species on changing landscapes*. Cambridge: Cambridge University Press.

Huxley, T. H., and H. N. Martin. 1875. *A course of practical instruction in elementary biology*. London: Macmillan and Co.

Isaacs, F. J., J. Hasty, C. R. Cantor, and J. J. Collins. 2003. Prediction and measurement of an autoregulatory genetic module. *Proceedings of the National Academy of Sciences USA* 100:7714-7719.

Istrail, S., and E. H. Davidson. 2005. Logic functions of the genomic *cis*-regulatory code. *Proceedings of the National Academy of Sciences USA* 102:4954-4959.

Jablonka, E., and M. Lamb. 1995. *Epigenetic inheritance and evolution: The Lamarckian dimension.* Oxford: Oxford University Press.

Jablonka, E., and M. Lamb. 2005. *Evolution in four dimensions: Genetic, epigenetic, behavioral, and symbolic variation in the history of life.* Cambridge, MA: MIT Press.

Jablonski, D. 1995. Extinctions in the fossil record. In *Extinction rates*, edited by J. H. Lawton and R. M. May. Oxford: Oxford University Press. Pp. 25-44.

Jablonski, D., K. Roy, and J. W. Valentine. 2006. Out of the tropics: Evolutionary dynamics of the latitudinal diversity gradient. *Science* 314:102-106.

Jacob, F. 1977. Evolution and tinkering. *Science* 196:1161-1166.

Jacob, F., and J. Monod. 1961. Genetic regulatory mechanisms in the synthesis of proteins. *Journal of Molecular Biology* 3:318-356.

Jahn, R., and R. H. Scheller. 2006. SNAREs-engines for membrane fusion. *Nature Reviews Molecular Cell Biology* 7:631-643.

Janzen, D. H. 1967. Why mountain passes are higher in the tropics. *The American Naturalist* 101:233-249.

Janzen, D. H. 1970. Herbivores and the number of tree species in tropical forests. *The American Naturalist* 104:501-528.

Janzen, F. 1994. Climate change and temperature-dependent sex determination in reptiles. *Proceedings of the National Academy of Sciences USA* 91:7487-7490.

Janzen, F., and J. Krenz. 2004. Phylogenetics: Which was first, TSD or GSD? In *Temperature-dependent sex determination in vertebrates*, edited by N. Valenzuela and V. Lance. Washington, DC: Smithsonian Books.

Jeong, H., B. Tombor, R. Albert, Z. N. Oltvai, and A.-L. Barabási. 2000. The large-scale organization of metabolic networks. *Nature* 407:651-654.

Jiang, R., Z. Tu, T. Chen, and F. Sun. 2006. Network motif identification in stochastic networks. *Proceedings of the National Academy of Sciences USA* 103:9404-9409.

Joshi, H. M., and F. R. Tabita. 1996. A global two-way component signal transduction system that integrates the control of photosynthesis, carbon dioxide assimilation and nitrogen fixation. *Proceedings of the National Academy of Sciences USA* 93:14515-14520.

Julius, D., and A. I. Basbaum. 2001. Molecular mechanisms of nocioception. *Nature* 413:203-210.

Jurka, J., V. V. Kapitonov, O. Kohany, and M. V. Jurka. 2007. Repetitive sequences in complex genomes: Structure and evolution. *Annual Review of Genomics and Human Genetics* 8:241-259.

Juroszek, P., and R. Gerhards. 2004. Photocontrol of weeds. *Journal of Agronomy and Crop Science* 190:402-415.

Kandel, E. R. 2001. The molecular biology of memory storage: A dialogue between genes and synapses. *Science* 294:1030-1038.

Kaplinsky, N. J., D. M. Braun, J. Penterman, S. A. Goff, and M. Freeling. 2002. Utility and distribution of conserved noncoding sequences in the grasses. *Proceedings of the National Academy of Sciences USA* 99:6147-6151.

Karban, R., C. A. Black, and S. A. Weinbaum. 2000. How 17-year cicadas keep track of time. *Ecology Letters* 3:253-256.

Keen, E. C., and A. J. Hudspeth. 2006. Transfer characteristics of the hair cell's afferent synapse. *Proceedings of the National Academy of Sciences USA* 103:5537-5542.

Keller, E. F. 2000. Models of and models for: Theory and practice in contemporary biology. *Philosophy of Science* 67(Proceedings):S72-86.

Kelly, L. 1956. A new interpretation of information rate. *AT&T Technical Journal* 35:917-926.

Kim, J. 2001a. After the molecular evolution revolution. *Evolution* 55:2620-2622.

Kim, J. 2001b. Descartes' fly: Geometry of genomic annotation. *Functional and Integrated Genomics* 1:241-249.

Kimura, M. 1961. Natural selection as a process of accumulating genetic information in adaptive evolution. *Genetical Research* 2:127-140.

Kimura, M. 1968. Evolutionary rate at the molecular level. *Nature* 217:624-626.

Kirkpatrick, M., and D. Lofsvold. 1992. Measuring selection and constraint in the evolution of growth. *Evolution* 46:954-971.

Kobayashi, M., T. Irino, and W. Sweldens. 2001. Multiscale computing. *Proceedings of the National Academy of Sciences USA* 98:12344-12345.

Koelle, K., S. Cobey, B. Grenfell, and M. Pascual. 2006. Epochal evolution shapes the phylodynamics of interpandemic influenza A (H3N2) in humans. *Science* 314:1898-1903.

Kolmogorov, A. N. 1965. Three approaches to the quantitative definition of information. *Problems of Information and Transmission* 1:1-7.

Korb, J., and K. E. Linsenmair. 1999. The architecture of termite mounds: A result of a trade-off between thermoregulation and gas exchange? *Behavioral Ecology* 10:312-316.

Korkin, D., F. P. Davis, F. Alber, T. Luong, M. Y. Shen, V. Lucic, M. B. Kennedy, and A. Sali. 2006. Structural modeling of protein interactions by analogy: Application to PSD-95. *Public Library of Science Computational Biology* 2:e153.

Korobkova, E., T. Emonet, J. M. G. Vilar, T. S. Shimizu, and P. Cluzel. 2004. From molecular noise to behavioural variability in a single bacterium. *Nature* 428:574-578.

Koropatnick, T. A., J. T. Engle, M. A. Apicella, E. V. Stabb, W. E. Goldman, and M. J. McFall-Ngai. 2004. Microbial factor-mediated development in a host-bacterial mutualism. *Science* 306:1186-1188.

Kozlov, A. S., T. Risler, and A. J. Hudspeth. 2007. Coherent motion of stereocilia assures the concerted gating of hair-cell transduction channels. *Nature Neuroscience* 10:87-92.

Kussell, E., and S. Leibler. 2005. Phenotypic diversity, population growth, and information in fluctuating environments. *Science* 309:2075-2078.

Lande, R. 1979. Quantitative genetic analysis of multivariate evolution, applied to brain:body size allometry. *Evolution* 33:402-416.

Lande, R., S. Engen, and B.-E. Saether. 2003. *Stochastic population dynamics in ecology and conservation.* Oxford: Oxford University Press.

Latham, R. E., and R. E. Ricklefs. 1993. Global patterns of tree species richness in moist forests: Energy-diversity theory does not account for variation in species richness. *Oikos* 67:325-333.

Levi, R., and J. M. Camhi. 2000. Wind direction coding in the cockroach escape response: Winner does not take all. *Journal of Neuroscience* 20:3814-3821.

Levine, M., and E. H. Davidson. 2005. Gene regulatory networks for development. *Proceedings of the National Academy of Sciences USA* 102:4936-4942.

Levins, R. 1966. The strategy of model building in population biology. *American Scientist* 54:421-431.

Levins, R. 1968. *Evolution in changing environments: Some theoretical explorations.* Princeton, NJ: Princeton University Press.

Lewen, G. D., W. Bialek, and R. R. de Ruyter van Steveninck. 2001. Neural coding of naturalistic motion stimuli. *Network* 12:317-329.

Lewis, J. E., and L. Maler. 2001. Neuronal population codes and the perception of object distance in weakly electric fish. *Journal of Neuroscience* 21:2842-2850.

Lewis, M. R., M. E. Carr, G. C. Feldman, W. Esaias, and C. McClain. 1990. Influence of penetrating solar radiation on the heat budget of the equatorial Pacific Ocean. *Nature* 347:543-545.

Lewontin, R. C. 1974. *The genetic basis of evolutionary change.* New York and London: Columbia University Press.

Ley, R. E., P. J. Turnbaugh, S. Klein, and J. I. Gordon. 2006. Microbial ecology: Human gut microbes associated with obesity. *Nature* 444:1022-1023.

Li, M., X. Chen, X. Li, B. Ma, and P. M. B. Vitanyi. 2004. The similarity metric. *IEEE Transactions on Information Theory* 50:3250-3264.

Lisman, J. E. 1985. A mechanism for memory storage insensitive to molecular turnover: A bistable autophosphorylating kinase. *Proceedings of the National Academy of Sciences USA* 82:3055-3057.

Lisman, J. E., and A. M. Zhabotinsky. 2001. A model of synaptic memory: A CaMKII/PP1 switch that potentiates transmission by organizing an AMPA receptor anchoring assembly. *Neuron* 31:191-201.

Livnat, A., and N. Pippenger. 2006. An optimal brain can be composed of conflicting agents. *Proceedings of the National Academy of Sciences USA* 103:3198-3202.

Lloyd, E. S. 1988. *The structure and confirmation of evolutionary theory.* New York: Greenwood Press.

Lo, C. C., and X. J. Wang. 2006. Cortico-basal ganglia circuit mechanism for a decision threshold in reaction time tasks. *Nature Neuroscience* 9:956-963.

Lopez-Schier, H., and A. J. Hudspeth. 2006. A two-step mechanism underlies the planar polarization of regenerating sensory hair cells. *Proceedings of the National Academy of Sciences USA* 103:18615-18620.

López-Urrutia, A., E. San Martín, R. P. Harris, and X. Irigoien. 2006. Scaling the metabolic balance of the oceans. *Proceedings of the National Academy of Sciences USA* 103:8739-8744.

Lynch, M., and R. Lande. 1993. Evolution and extinction in response to environmental change. In *Biotic interactions and global change,* edited by P. M. Kareiva, J. G. Kingsolver, and R. B. Huey. Sunderland: Sinauer.

Ma, W. J., J. M. Beck, P. E. Latham, and A. Pouget. 2006. Bayesian inference with probabilistic population codes. *Nature Neuroscience* 9:1432-1438.

MacArthur, R. H., and E. O. Wilson. 1967. *The theory of island biogeography.* Princeton, NJ: Princeton University Press.

Mackay, A. L. 1991. *A dictionary of scientific quotations.* London: Institute of Physics Publishing.

Magwene, P. M. 2001. New tools for studying integration and modularity. *Evolution* 55:1734-1745.

Marder, E., and D. Bucher. 2001. Central pattern generators and the control of rhythmic movements. *Current Biology* 11:R986-R996.

Marder, E., and R. L. Calabrese. 1996. Principles of rhythmic motor pattern generation. *Physiological Reviews* 76:687-717.

Mason, A. C., M. L. Oshinsky, and R. R. Hoy. 2001. Hyperacute directional hearing in a microscale auditory system. *Nature* 410:644-645.

Marzluff, J. M., B. Heinrich, and C. S. Marzluff. 1996. Raven roosts are mobile information centres. *Animal Behaviour* 51:89-103.

Matsuoka, K., R. Schekman, L. Orci, and J. E. Heuser. 2001. Surface structure of the COPII-coated vesicle. *Proceedings of the National Academy of Sciences USA* 98:13705-13709.

Matzke, M. A., and J. A. Birchler. 2005. RNAi-mediated pathways in the nucleus. *Nature Reviews Genetics* 6:24-35.

Maynard Smith, J., and E. Szathmáry. 1995. *The major transitions in evolution.* Oxford: W. H. Freeman/Spektrum.

Melquist, S., B. Luff, and J. Bender. 1999. *Arabidopsis* PAI gene arrangements, cytosine methylation and expression. *Genetics* 153:401-413.

Menashe, I., O. Man, D. Lancet, and Y. Gilad. 2003. Different noses for different people. *Nature Genetics* 34:143-144.

Metzner, W. 1993. The jamming avoidance response in Eigenmannia is controlled by two separate motor pathways. *Journal of Neuroscience* 13:1862-1878.

Miles, E. W., S. Rhee, and D. R. Davies. 1999. The molecular basis of substrate channeling. *Journal of Biological Chemistry* 274:12193-12196.

Miller, P., A. M. Zhabotinsky, J. E. Lisman, and X. J. Wang. 2005. The stability of a stochastic CaMKII switch: Dependence on the number of enzyme molecules and protein turnover. *Public Library of Science Biology* 3:e107.

Mirkin, S. M. 2006. DNA structures, repeat expansions and human hereditary disorders. *Current Opinion in Structural Biology* 16:351-358.

Moorcroft, P. R. 2006. How close are we to a predictive science of the biosphere? *Trends in Ecology and Evolution* 21:400-440.

Moret, B. M. E. 2005. Computational challenges from the Tree of Life in *Proceedings of the 7th Workshop on Algorithm Engineering and Experiments,* pp. 3-16. Philadelphia, PA: SIAM Press.

Morgante, M. 2006. Plant genome organization and diversity: The year of the junk! *Current Opinion in Biotechnology* 17:168-173.

Nakajima, M., K. Imai, I. Hiroshi, T. Nishiwaki, Y. Murayama, H. Iwasaki, T. Oyama, and T. Kondo. 2005. Reconstitution of circadian oscillation of cyanobacterial KaiC phosphorylation in vitro. *Science* 308:414-415.

National Research Council (NRC). 1989. *Opportunities in biology.* Washington, DC: National Academy Press.

NRC. 1996. *National Science Education Standards.* Washington, DC: National Academy Press.

NRC. 1998. *Teaching about evolution and the nature of science.* Washington, DC: National Academy Press.

NRC. 2007. *The new science of metagenomics.* Washington, DC: The National Academies Press.

National Science Board. 2007. *Enhancing support for transformative research at the National Science Foundation.* Arlington, VA: National Science Foundation.

NSF (National Science Foundation). 1998. *Shaping the future, volume II: Perspectives on undergraduate education in science, mathematics, engineering, and technology.* Arlington, VA: National Science Foundation.

Nealson, K. H., and R. Rye. 2005. Evolution of metabolism. In *Treatise on geochemistry: Volume 8, Biogeochemistry,* edited by W. H. Schlesinger and executive editors H. D. Holland and K. K. Turekian. Amsterdam: Elsevier. Pp. 41-61.

Nesse, R. M., S. C. Stearns, and G. S. Ommen. 2006. Medicine needs evolution. *Science* 311:1071.

Neuweiler, G. 1990. Auditory adaptations for prey capture in echolocating bats. *Physiological Reviews* 70:615-641.

Nisbet, E. G., and C. M. R. Fowler. 2005. The early history of life. In *Treatise on geochemistry: Volume 8, Biogeochemistry,* edited by W. H. Schlesinger. Amsterdam: Elsevier. Pp. 1-39.

Novotny, V., P. Drozd, S. E. Miller, M. Kulfan, M. Janda, Y. Basset, and G. D. Weiblen. 2006. Why are there so many species of herbivorous insects in tropical rainforests? *Science* 313:1115-1118.

Nyquist, H., and R. V. L. Hartley. 1928. Transmission of information. *Bell System Technical Journal.* Pp. 535-563.

O'Connell-Rodwell, C., J. Wood, C. Kinzley, T. Rodwell, J. Poole, and S. Puria. 2007. Wild African elephants (*Loxodonta africana*) discriminate subtle differences between conspecific seismic cues. *Journal of the Acoustical Society of America* 122:823-830.

Odling-Smee, F. J., K. N. Laland, and M. W. Feldman. 2003. *Niche construction: The neglected process in evolution.* Princeton, NJ: Princeton University Press.

Ohno, S. 1967. *Sex chromosomes and sex-linked genes.* Berlin-Heidelberg-New York: Springer.

Orr, H. A. 2000. Adaptation and the cost of complexity. *Evolution* 54:13-20.

Ovadi, J., and P. A. Srere. 1992. Channel your energies. *Trends in Biochemical Sciences* 17:445-447.

Oyama S., P. E. Griffiths, and R. D. Gray. 2001. *Cycles of contingency: Developmental systems and evolution, life and mind.* Cambridge, MA: MIT Press.

Park, T. 1941. The laboratory population as a test of a comprehensive ecological system. *Quarterly Review of Biology* 16:274-293,440-461.

Parmesan, C. 2006. Ecological and evolutionary responses to recent climate change. *Annual Review of Ecology Evolution and Systematics* 37:637-669.

Parmesan, C., S. Gaines, L. Gonzalez, D. M. Kaufman, J. Kingsolver, A. Townsend Peterson, and R. Sagarin. 2005. Empirical perspectives on species borders: From traditional biogeography to global change. *Oikos* 108:58-75.

Pasquali, S., H. H. Gan, and T. Schlick. 2005. Modular RNA architecture revealed by computational analysis of existing pseudoknots and ribosomal RNAs. *Nucleic Acids Research* 33:1384-1398.

Pedraza, J. M., and A. van Oudenaarden. 2005. Noise propagation in gene networks. *Science* 307:1965-1969.

Pereira-Leal, J. B., E. D. Levy, C. Kamp, and S. A. Teichmann. 2007. Evolution of protein complexes by duplication of homomeric interactions. *Genome Biology* 8:R51.

Piatigorsky, J. 2007. *Gene sharing and evolution.* Cambridge, MA: Harvard University Press.

Pimm, S. L., J. H. Lawton, and J. E. Cohen. 1991. Food web patterns and their consequences. *Nature* 350:669-674.

Ponting, C. P., and R. B. Russell. 1995. Swaposins: Circular permutations within genes encoding saposin homologs. *Trends in Biochemical Sciences* 20:179-180.

Popper, K. 1959. *The logic of scientific discovery.* New York: Basic Books.

Qin, H., H. H. S. Lu, W. B. Wu, and W. H. Li. 2003. Evolution of the yeast protein interaction network. *Proceedings of the National Academy of Sciences USA* 100:12820-12824.

Raff, R. A. 1996. *The shape of life: Genes, development, and the evolution of animal form.* Chicago, IL: University of Chicago Press.

Raghoebarsing, A. A., A. Pol, K. T. van de Pas-Schoonen, A. J. P. Smolders, K. F. Ettwig, W. I. C. Rijpstra, S. Schouten, J. S. Sinninghe Damsté, H. J. M. Op den Camp, M. S. M. Jetten, and M. Strous. 2006. A microbial consortium couples anaerobic methane oxidation to denitrification. *Nature* 440:918-921.

Rainey, H. J., K. Zuberbuhler, and P. J. Slater. 2004. Hornbills can distinguish between primate alarm calls. *Proceedings of the Royal Society B—Biological Sciences* 271:755-759.

Rakic, P., J. P. Bourgeois, and P. S. Goldman-Rakic. 1994. Synaptic development of the cerebral cortex: Implications for learning, memory, and mental illness. *Progress in Brain Research* 102:227-243.

Ramsey, I. S., M. Delling, and D. E. Clapham. 2006. An introduction to TRP channels. *Annual Review of Physiology* 68:619-647.

Rana, T. M. 2007. Illuminating the silence: Understanding the structure and function of small RNAs. *Nature Reviews Molecular Cell Biology* 8:23-36.

Rassoulzadegan, M., V. Grandjean, P. Gounon, S. Vincent, I. Gillot, and F. Cuzin. 2006. RNA-mediated non-mendelian inheritance of an epigenetic change in the mouse. *Nature* 441:469-474.

Rausher, M. D. 2001. Co-evolution and plant resistance to natural enemies. *Nature* 411:857-864.

Ravasz, E., A. L. Somera, D. A. Mongru, Z. N. Oltvai, and A.-L. Barabási. 2002. Hierarchical organization of modularity in metabolic networks. *Science* 297:1551-1555.

Redies, C., and L. Puelles. 2001. Modularity in vertebrate brain development and evolution. *Bioessays* 23:1100-1111.

Redon, R., S. Ishikawa, K. R. Fitch, L. Feuk, G. H. Perry, T. D. Andrews, H. Fiegler, M. H. Shapero, A. R. Carson, W. Chen, E. K. Cho, S. Dallaire, J. L. Freeman, J. R. Gonzalez, M. Gratacos, J. Huang, D. Kalaitzopoulos, D. Komura, J. R. MacDonald, C. R. Marshall, R. Mei, L. Montgomery, K. Nishimura, K. Okamura, F. Shen, M. J. Somerville, J. Tchinda, A. Valsesia, C. Woodwark, F. T. Yang, J. J. Zhang, T. Zerjal, J. Zhang, L. Armengol, D. F. Conrad, X. Estivill, C. Tyler-Smith, N. P. Carter, H. Aburatani, C. Lee, K. W. Jones, S. W. Scherer, and M. E. Hurles. 2006. Global variation in copy number in the human genome. *Nature* 444:444-454.

Ricklefs, R. E. 2004. A comprehensive framework for global patterns in biodiversity. *Ecology Letters* 7:1-15.

Ricklefs, R. E., and R. E. Latham. 1992. Intercontinental correlation of geographical ranges suggests stasis in ecological traits of relict genera of temperate perennial herbs. *American Naturalist* 139:1305-1321.

Ricklefs, R. E., and D. Schluter. 1993. *Species diversity in ecological communities: Historical and geographical perspectives.* Chicago, IL: University of Chicago Press.

Rosenfeld, N., J. W. Young, U. Alon, P. S. Swain, and M. B. Elowitz. 2005. Gene regulation at the single-cell level. *Science* 307:1962-1965.

Rosenzweig, M. R. 1975. On continental steady states of species diversity. In *Ecology and evolution of communities*, edited by M. L. Cody and J. M. Diamond. Cambridge: Harvard University Press. Pp. 121-140.

Roughgarden, J. 1979. *Theory of population genetics and evolutionary ecology: An introduction.* New York: Macmillan.

Roughgarden, J., M. Oishi, and E. Akcay. 2006. Reproductive social behavior: Cooperative games to replace sexual selection. *Science* 311:965-969.

Saier, M. H. 2000. A functional-phylogenetic classification system for transmembrane solute transporters. *Microbiology and Molecular Biology Reviews* 64:354-411.

Samoilov, M. S., G. Price, and A. P. Arkin. 2006. From fluctuations to phenotypes: The physiology of noise. *Science Signal Transduction Knowledge Environment* 2006:re17.

Saxe, J. G. 1878. *The Blind Men and the Elephant* in Linton's "Poetry of America." London: G. Bell.

Schaschi, H., P. Wandeler, F. Suchentruk, G. Oeber-Ruff, and S. J. Goodman. 2006. Selection and recombination drive the evolution of MHC class II DRB diversity in ungulates. *Heredity* 97:427-437.

Scheffer, M., S. Carpenter, and B. de Young. 2005. Cascading effects of overfishing marine systems. *Trends in Ecology and Evolution* 20:579-581.

Schlosser, G., and G. P. Wagner, Eds. 2004. *Modularity in development and evolution.* Chicago, IL: University of Chicago Press.

Scopel, A. L., C. L. Ballaré, and R. A. Sánchez. 1991. Induction of extreme light sensitivity in buried weed seeds and its role in the perception of soil cultivations. *Plant, Cell & Environment* 14:501-508.

Segrè, D., A. Deluna, G. M. Church, and R. Kishony. 2005. Modular epistasis in yeast metabolism. *Nature Genetics* 37:77-83.

Seung, H. S., D. D. Lee, B. Y. Reis, and D. W. Tank. 2000. Stability of the memory of eye position in a recurrent network of conductance-based model neurons. *Neuron* 26:259-271.

Shannon, C. E. 1948. A mathematical theory of communication. *AT&T Technical Journal* 27:279-423, 623-656.

Shannon, C. E., and W. Weaver. 1949. *The mathematical theory of communication.* Urbana: University of Illinois Press.

Shifman, J. M., M. H. Choi, S. Mihalas, S. L. Mayo, and M. B. Kennedy. 2006. Ca2+/calmodulin-dependent protein kinase II (CaMKII) is activated by calmodulin with two bound calciums. *Proceedings of the National Academy of Sciences USA* 103:13968-13973.

Siepel A., G. Bejerano, J. S. Pederson, A. S. Hinrichs, M. Hou, K. Rosenbloom, H. Clawson, J. Spieth, L. W. Hillier, S. Richards, G. M. Weinstock, R. K. Wilson, R. A. Gibbs, W. J. Kent, W. Miller, and D. Haussler. 2005. Evolutionarily conserved elements in vertebrate, insect, worm and yeast genomes. *Genome Research* 15:1034-1050.

Simmons, J. A., N. Neretti, N. Intrator, R. A. Altes, M. J. Ferragamo, and M. I. Sanderson. 2004. Delay accuracy in bat sonar is related to the reciprocal of normalized echo bandwidth, or Q. *Proceedings of the National Academy of Sciences USA* 101:3638-3643.

Sivan, E., and N. Kopell. 2006. Oscillations and slow patterning in the antennal lobe. *Journal of Computational Neuroscience* 20:85-96.

Smith, D. J., A. S. Lapedes, J. C. de Jong, T. M. Bestebroer, G. F. Rimmelzwaan, A. D. M. E. Osterhaus, and R. A. M. Fouchier. 2004. Mapping the antigenic and genetic evolution of influenza virus. *Science* 305:371-376.

Smotherman, M., and W. Metzner. 2003. Effects of echo intensity on Doppler-shift compensation behavior in horseshoe bats. *Journal of Neurophysiology* 89:814-821.

Snel, B., and M. A. Huynen. 2004. Quantifying modularity in the evolution of biomolecular systems. *Genome Research* 14:391-397.

Solomonoff, R. J. 1964. A formal theory of inductive inference, parts 1 & 2. *Information and Control* 7:1-22, 224-254.

Soppe, W. J., S. E. Jacobsen, C. Alonso-Blanco, J. P. Jackson, T. Kakutani, M. Koornneef, and A. J. Peeters. 2000. The late flowering phenotype of fwa mutants is caused by gain-of-function epigenetic alleles of a homeodomain gene. *Molecular Cell* 6:791-802.

Sosinsky, G. E., T. J. Deerinck, R. Greco, C. H. Buitenhuys, T. M. Bartol, and M. H. Ellisman. 2005. Development of a model for microphysiological simulations: Small nodes of ranvier from peripheral nerves of mice reconstructed by electron tomography. *Neuroinformatics* 3:133-162.

Sowers, T., and M. Bender. 1995. Climate records covering the last deglaciation. *Science* 269:210-214.

Stahl, E. A., G. Dwyer, R. Mauricioa, M. Kreitman, and J. Bergelson. 1999. Dynamics of disease resistance polymorphism at the *Rpm1* locus of *Arabidopsis*. *Nature* 400:667-671.

Stanford Linear Accelerator Center (SLAC). 2007. *Is the standard model a theory or a model?* Accessed on May 10, 2007 at http://www2.slac.stanford.edu/vvc/theory/modeltheory.html.

Stebbins, G. L. 1974. *Flowering plants: Evolution above the species level.* Cambridge, MA: Harvard University Press.

Stephens, J. C., J. A. Schneider, D. A. Tanguay, J. Choi, T. Acharya, S. E. Stanley, R. Jiang, C. J. Messer, A. Chew, J. H. Han, J. Duan, J. L. Carr, M. S. Lee, B. Koshy, A. M. Kumar, G. Zhang, W. R. Newell, A. Windemuth, C. Xu, T. S. Kalbfleisch, S. L. Shaner, K. Arnold, V. Schulz, C. M. Drysdale, K. Nandabalan, R. S. Judson, G. Ruano, and G. F. Vovis. 2001. Haplotype variation and linkage disequilibrium in 313 human genes. *Science* 293:489-493.

Sterner, R. W., and J. J. Elser. 2002. *Ecological stoichiometry*. Princeton, NJ: Princeton University Press.

Stolc, V., M. P. Samanta, W. Tongprasit, H. Sethi, S. Liang, D. C. Nelson, A. Hegeman, C. Nelson, D. Rancour, S. Bednarek, E. L. Ulrich, Q. Zhao, R. L. Wrobel, C. S. Newman, B. G. Fox, G. N. Phillips, J. L. Markley, and M. R. Sussman. 2005. Identification of transcribed sequences in *Arabidopsis thaliana* by using high-resolution genome tiling arrays. *Proceedings of the National Academy of Sciences USA* 22:102:4453-4458.

Stoleru, D., Y. Peng, J. Agosto, and M. Rosbash. 2004. Coupled oscillators control morning and evening locomotor behaviour of Drosophila. *Nature* 431:862-868.

Stoleru, D., Y. Peng, P. Nawathean, and M. Rosbash. 2005. A resetting signal between *Drosophila* pacemakers synchronizes morning and evening activity. *Nature* 438:238-242.

Sung, S., and R. M. Amasino. 2005. Remembering winter: Toward a molecular understanding of vernalization. *Annual Review of Plant Biology* 56:491-508.

Tan, K., T. Shlomi, H. Feizi, T. Ideker, and R. Sharan. 2007. Transcriptional regulation of protein complexes within and across species. *Proceedings of the National Academy of Sciences USA* 104:1283-1288.

Tank, D. W., and J. J. Hopfield. 1987. Collective computation in neuronlike circuits. *Scientific American* 257:104-114.

Teller, P. 2001. Twilight of the perfect model model. *Erkenntnis* 55:393-415.

Tettelin, H., V. Masignani, M. J. Cieslewicz, C. Donati, D. Medini, N. L. Ward, S. V. Angiuoli, J. Crabtree, A. L. Jones, A. S. Durkin, R. T. Deboy, T. M. Davidsen, M. Mora, M. Scarselli, I. Margarit y Ros, J. D. Peterson, C. R. Hauser, J. P Sundaram, W. C. Neslon, R. Madupu, L. M. Brinkac, R. J. Dodson, M. J. Rosovitz, S. A. Sullivan, S. C. Daugherty, D. H. Haft, J. Selengut, M. L. Gwinn, L. Zhou, N. Zafar, H. Khouri, D. Radune, G. Dimitrov, K. Watkins, K. J. O'Connor, S. Smith, T. R. Utterback, O. White, C. E. Rubens, G. Grandi, L. C. Madoff, D. L. Kasper, J. L. Telford, M. R. Wessels, R. Rappuoli, and C. M. Fraser. 2005. Genome analysis of multiple pathogenic isolates of *Streptococcus agalactiae*: Implications for the microbial "pan-genome." *Proceedings of the National Academy of Sciences USA* 102:13950-13955.

Theodoropoulus, C., Y.-H. Qian, and I. G. Kevrekidis. 2000. Coarse stability and bifurcation analysis using time-steppers: A reaction-diffusion example. *Proceedings of the National Academy of Sciences USA* 97:9840-9843.

Thomson, E. E., and W. B. Kristan. 2005. Quantifying stimulus discriminability: A comparison of information theory and ideal observer analysis. *Neural Computation* 17:741-778.

Tree of Life Web Project. 1997. Life on Earth. Version 01 January 1997 (temporary). Accessed on January 15, 2007 at http://tolweb.org/Life_on_Earth/1/1997.01.01 in The Tree of Life Web Project, http://tolweb.org/.

Tringe, S. G., C. von Mering, A. Kobayashi, A. A. Salamanov, K. Chang, H. Hwai, M. Podar, J. M. Short, E. J. Mahur, J. C. Detter, P. Bork, P. Hugenholtz, and E. M. Rubin. 2005. Comparative metagenomics of microbial communities. *Science* 308:554-557.

Tu, B. P., A. Kudlicki, M. Rowicka, and S. L. McKnight. 2005. Logic of the yeast metabolic cycle: Temporal compartmentalization of cellular processes. *Science* 310:1152-1158.

Turnbaugh, P. J., R. E. Ley, M. A. Mahowald, V. Magrini, E. R. Mardis, and J. I. Gordon. 2006. An obesity-associated gut microbiome with increased capacity for energy harvest. *Nature* 444:1027-1031.

Turner, B. M. 2001. *Chromatin and gene regulation: Molecular mechanisms in epigenetics*. Oxford: Blackwell Science.

Valenzuela, N. 2004. Conclusions: Missing links and future directions. In *Temperature-dependent sex determination in vertebrates*, edited by N. Valenzuela and V. Lance. Washington, DC: Smithsonian Books. Pp. 157-160.

Valenzuela, N., A. LeClere, and T. Shikano. 2006. Comparative gene expression of steroidogenic factor 1 in *Chrysemys picta* and *Apalone mutica* turtles with temperature-dependent and genotypic sex determination. *Evolution and Development* 8:424-432.

Van Dover, C. L. 2000. *The ecology of deep-sea hydrothermal vents*. Princeton, NJ: Princeton University Press. P. 424.

Venter, J. C., K. Remington, J. F. Heidelberg, A. L. Halpern, D. Rusch, J. A. Eisen, D. Wu, I. Paulsen, K. E. Nelson, W. Nelson, D. E. Fouts, S. Levy, A. H. Knap, M. W. Lomas, K. Nealson, O. White, J. Peterson, J. Hoffman, R. Parsons, H. Baden-Tillson, C. Pfannkock, Y.-H. Rogers, and H. O. Smith. 2004. Environmental genome shotgun sequencing of the Sargasso Sea. *Science* 304:66-74.

Von Dassow, G., and E. Meir. 2004. Exploring modularity with dynamical models of gene networks. In *Modularity in Development and Evolution*, edited by G. Schlosser and G. P. Wagner. Chicago, IL: University of Chicago Press. Pp. 244-287.

Wagner, G. P. 1988. The influence of variation and of developmental constraints on the rate of multivariate phenotypic evolution. *Journal of Evolutionary Biology* 1:45-66.

Wagner, A. 2002. Selection and gene duplication: A view from the genome. *Genome Biology* 3:1012.1-1012.3.

Wagner, G. P. 1996. Homologues, natural kinds and the evolution of modularity. *American Zoologist* 36:36-43.

Wagner, G. P., G. Booth, and H. Bagheri-Chaichian. 1997. A population genetic theory of canalization. *Evolution* 51:329-347.

Wagner, G. P., and J. A. Gauthier. 1999. 1,2,3=2,3,4: A solution to the problem of the homology of the digits in the avian hand. *Proceedings of the National Academy of Sciences USA* 96:5111-5116.

Waterston, R. H., K. Lindblad-Toh, E. Bimey, J. Rogers, J. F. Abril, P. Agarwal, R. Agarwala, R. Ainscough, M. Alexandersson, P. An, S. E. Antonarakis, J. Attwood, R. Baertsch, J. Bailey, K. Barlow, S. Beck, E. Berry, B. Birren, T. Bloom, P. Bork, M. Botcherby, N. Bray, M. R. Brent, D. G. Brown, S. D. Brown, C. Bult, J. Burton, J. Butler, R. D. Campbell, P. Carninci, S. Cawley, F. Chiaromonte, A. T. Chinwalla, D. M. Church, M. Clamp, C. Clee, F. S. Collins, L. L. Cook, R. R. Copley, A. Coulson, O. Couronne, J. Cuff, V. Curwen, et al. 2002. Initial sequencing and comparative analysis of the mouse genome. *Nature* 420:520-562.

Watson, J. D., and F. H. Crick. 1953 Molecular structure of nucleic acids. *Nature* 171:737-738.

Webster, A. J., R. J. H. Payne, and M. Pagel. 2003. Molecular phylogenies link rates of evolution and speciation. *Science* 301:478-478.

Weismann, A. 1904. *The Evolution theory* (translation: J. A. and M. R. Thomson). London: Edward Arnold.

West, G. B., J. H. Brown, and B. J. Enquist. 1997. A general model for the origin of allometric scaling laws in biology. *Science* 276:122-126.

Wheeler, W. M. 1923. The dry-rot of our academic biology. *Science* 57:61-71.

White, J. A., C. C. Chow, J. Ritt, C. Soto-Trevino, and N. Kopell. 1998. Synchronization and oscillatory dynamics in heterogeneous, mutually inhibited neurons. *Journal of Computational Neuroscience* 5:5-16.

Wiener, N. 1948. *Cybernetics: Or the control and communication in the animal and the machine*. Cambridge: MIT Press.

Wiens, J. J., and M. J. Donoghue. 2004. Historical biogeography, ecology and species richness. *Trends in Ecology & Evolution* 19:639-644.

Wiens, J. J., and C. H. Graham. 2005. Niche conservatism: Integrating evolution, ecology, and conservation biology. *Annual Review of Ecology Evolution and Systematics* 36:519-539.

Wiens, J. J., C. H. Graham, D. S. Moen, S. A. Smith, and T. W. Reeder. 2006. Evolutionary and ecological causes of the latitudinal diversity gradient in hylid frogs: Treefrog trees unearth the roots of high tropical diversity. *American Naturalist* 168:579-596.

Williams, G. C. 1966. *Adaptation and natural selection: A critique of some current evolutionary thought.* Princeton, NJ: Princeton University Press.

Williams, J. W., B. N. Schumer, T. Webb III, P. J. Bartlein, and P. L. Leduc. 2004. Late-quarternary vegetation dynamics in North America: Scaling from taxa to biomes. *Ecological Monographs* 74:309-334.

Willig, M. R., D. M. Kaufman, and R. D. Stevens. 2003. Latitudinal gradients of biodiversity: Pattern, process, scale, and synthesis. *Annual Review of Ecology Evolution and Systematics* 34:273-309.

Willingham, E., R. Baldwin, J. Skipper, and D. Crews. 2000. Aromatase activity during embryogenesis in the brain and adrenal–kidney–gonad of the red-eared slider turtle, a species with temperature-dependent sex determination. *General and Comparative Endocrinology* 119:202–207.

Wimsatt, W. C. 1987. False models as means to truer theories. In *Neutral models in biology,* edited by M. Nitecki and A. Hoffman. London: Oxford University Press. Pp. 23-55.

Wimsatt, W. C. 2007. *Re-engineering philosophy for limited beings: Piecewise approximations to reality.* Cambridge, MA: Harvard University Press.

Winther, R. G. 2001. Varieties of modules: Kinds, levels, origins, and behaviors. *Journal of Experimental Zoology* 291:116-129.

Woese, C. 1998. The universal ancestor. *Proceedings of the National Academy of Sciences USA* 95:6854-6859.

Woese, C. R., O. Kandler, and M. L. Wheelis. 1990. Towards a natural system of organisms: Proposal for the domains Archaea, Bacteria, and Eucarya. *Proceedings of the National Academy of Sciences USA* 87:4576-4579.

Wolfe, K. H., and W-H. Li. 2003. Molecular evolution meets the genomics revolution. *Nature Genetics* 33:255-265.

Wood, A. J., and R. J. Oakey. 2006. Genomic imprinting in mammals: Emerging themes and established theories. *Public Library of Science Genetics* 2:e147.

Worm, B., E. B. Barbier, N. Beaumont, J. E. Duffy, C. Folke, B. S. Halpern, J. B. C. Jackson, H. K. Lotze, F. Micheli, S. R. Palumbi, E. Sala, K. A. Selkoe, J. J. Stachowicz, and R. Watson. 2006. Impacts of biodiversity loss on ocean ecosystem services. *Science* 314:787-790.

Wright, J., R. E. Stone, and N. Brown. 1998. Communal roosts as structured information centres in the raven, *Corvus corax. Journal of Animal Ecology* 72:1003.

Yang, A. S. 2001. Modularity, evolvability, and adaptive radiations: A comparison of the hemi- and holometabolous insects. *Evolution and Development* 3:59-72.

Yeger-Lotem, E., S. Sattath, N. Kashtan, S. Itzkovitz, R. Milo, R. Y. Pinter, U. Alon, and H. Margalit. 2005. Network motifs in integrated cellular networks of transcription-regulation and protein-protein interaction. *Proceedings of the National Academy of Sciences USA* 101:5934-5939.

Yu, H., and M. Gerstein. 2006. Genomic analysis of the hierarchical structure of regulatory networks. *Proceedings of the National Academy of Sciences USA* 103:14724-14731.

Zakon, H. H., and K. D. Dunlap. 1999. Sex steroids and communication signals in electric fish: A tale of two species. *Brain, Behavior and Evolution* 54:61-69.

Zakon, H., J. Oestreich, S. Tallarovic, and F. Triefenbach. 2002. EOD modulations of brown ghost electric fish: JARs, chirps, rises, and dips. *Journal of Physiology—Paris* 96:451-458.

Zaratiegui, M., D. V. Irvine, and R. A. Martienssen. 2007. Noncoding RNAs and gene silencing. *Cell* 128:763-766.

Zhang, B., Q. Wang, and X. Pan. 2007. MicroRNAs and their regulatory roles in animals and plants. *Journal of Cellular Physiology* 210:279-289.

Zhang, J. 2003. Evolution by gene duplication: An update. *Trends in Ecology and Evolution* 18:292-298.

Zhang, J., Y. Zhange, and H. F. Rosenberg. 2002. Adaptive evolution of a duplicated pancreatic ribonuclease gene in a leaf-eating monkey. *Nature Genetics* 30:411-415.

Zhang, T., M. Breitbart, W. H. Lee, J. Q. Run, C. L. Wei, S. W. Soh, M. L. Hibberd, E. T. Liu, F. Rohwer, and Y. Ruan. 2006. RNA viral community in human feces: Prevalence of plant pathogenic viruses. *Public Library of Science Biology* 4:e3.

Appendix A

Statement of Task

The Board on Life Sciences will convene an ad hoc committee to identify and examine the concepts and theories that form the foundation for scientific advancement in various areas of biology, including (but not limited to) genes, cells, ecology, and evolution. It will assess which areas are "theory-rich" and which areas need stronger conceptual foundations for substantial advancement and make recommendations as to the best way to encourage creative, dynamic, and innovative research in biology. Building on these results, the study will identify major questions to be addressed by 21st-century biology. The project will focus on basic biology, but not on biomedical applications. Questions to be considered by the committee may include:

- What does it mean to think of biology as a theoretical science?
- Is there a basic set of theories and concepts that are understood by biologists in all subdisciplines?
- How do biological theories form the foundation for scientific advancement?
- Which areas of biology are "theory-rich" and which areas need stronger conceptual foundations for substantial advancement?
- What are the best ways to bring about advances in biology?
- What are the grand challenges in 21st-century biology?
- How can educators ensure that students understand the foundations of biology?

Appendix B

Biographical Sketches of Committee Members

David J. Galas (Chair) is the vice president and chief scientific officer of biological and life sciences at Battelle Memorial Institute and a professor at the Institute for Systems Biology. Prior to joining Battelle, he held a number of key positions, including chancellor, chief scientific officer, and Norris Professor of Applied Life Sciences, Keck Graduate Institute of Life Sciences (Claremont, California); president and chief scientific officer of Darwin Molecular and Chiroscience R&D, Inc.; director for health and environmental research, U.S. Department of Energy (DOE); professor of molecular biology, University of Southern California; senior staff scientist, Biomedical Division, Lawrence Livermore National Laboratory; and scientific adviser to the Defense Science Board. Dr. Galas was among the early leaders in the U.S. Human Genome Program, and from 1990 to 1993 he led the DOE component of this project. In recent years his focus has included the application of a variety of scientific disciplines to addressing challenging biological and medical problems, including Alzheimer's disease, osteoporosis, and improving diagnostic technologies. Dr. Galas received his Ph.D. in physics from the University of California, Davis-Livermore, and his undergraduate degree in physics from the University of California, Berkeley. He is the recipient of several awards, including the Smithsonian Institution Computer World Pioneer Award in 1999. He has served on many federal, university, and corporate boards and advisory committees, including a number of National Research Council committees and the Board on Life Science. He is a lifetime national associate of the National Academy of Sciences.

Carl T. Bergstrom is an associate professor in the Department of Biology at the University of Washington. As an evolutionary biologist, Dr. Bergstrom studies the role of information in biological systems at scales from intracellular control of gene expression to population-wide linguistic communication. Working in close collaboration with empirical and experimental researchers, Dr. Bergstrom's group approaches these problems using mathematical models and computer simulations. His recent projects include contributions to the game theory of communication, models of intracellular information processing, and work on how immune systems avoid subversion by pathogens. In a set of more applied endeavors, Dr. Bergstrom uses ecological and evolutionary theory to understand and control emerging infectious diseases, including antibiotic-resistant bacteria in hospitals and novel emerging pathogens such as SARS and avian influenza. A national leader in promoting public education about evolutionary biology, Dr. Bergstrom received his Ph.D. in theoretical population genetics from Stanford University in 1998. After a two-year postdoctoral fellowship at Emory University, where he studied the ecology and evolution of infectious diseases, Dr. Bergstrom joined the faculty at the University of Washington in 2001.

Vicki L. Chandler holds the Carl E. and Patricia Weiler Endowed Chair for Excellence in Agriculture and Life Sciences, regents' professor in the Department of Plant Sciences and Molecular and Cellular Biology, and the director of the BIO5 Institute at the University of Arizona. She received her B.A. from the University of California, Berkeley, and her Ph.D. from the University of California, San Francisco. Dr. Chandler has conducted pioneering research on the control of gene expression in plants and animals. She has received numerous honors and awards, including the Presidential Young Investigator Award, the Searle Scholar Award, the National Science Foundation (NSF) Faculty Award for Women Scientists and Engineers, and the NIH Director's Pioneer Award. She has served extensively on national advisory boards and panels for NSF, DOE, NIH, and HHMI, including the NSF Biological Directorate Advisory Committee from 2001 to 2004. She has chaired or cochaired national conferences for Keystone, FASEB, and the Gordon Research Conferences, serving on the GRC board of trustees and in 2001 as chair of the board. Dr. Chandler was elected to the International Society of Plant Molecular Biology Board of Directors for 1999-2003 and president of the American Society of Plant Biologists for 2001-2002. In 2007 she was elected to the Council of the National Academy of Sciences, to which she was elected a member in 2002.

Paul G. Falkowski is a professor of biochemistry and biophysics at the Institute of Marine and Coastal Sciences, Rutgers University. Some of his research interests include biogeochemical cycles, photosynthesis, biological oceanography, molecular biology, biochemistry and biophysics, physiological adaptation, plant physiology, evolution, mathematical modeling, and symbiosis. Dr. Falkowski is also the lead principal investigator in the Environmental Biophysics and Molecular Ecology program. That program focuses on molecular biology and biophysics to address key questions in biological oceanography and marine biology. The program also provides a laboratory in the Institute of Marine and Coastal Sciences at Rutgers University that addresses the application of similar techniques to primary production, nitrogen fixation, and other rate-determining processes in aquatic and terrestrial ecosystems. Dr. Falkowski has received many awards; his most recent include the Board of Trustees Award for Excellence in Research, Rutgers University (2000); the Vernadsky Medal, European Geosciences Union (2005); and the Board of Governors Professor, Rutgers University (2005). He has also received numerous grants, some from NASA, NSF, DOD, DOE, and the Moore Foundation. Dr. Falkowski received his Ph.D. in biology from the University of British Columbia. He was elected to the National Academy of Sciences in 2007.

Douglas J. Futuyma is a distinguished professor of ecology and evolution at the State University of New York at Stony Brook. He received his M.S. and Ph.D. from the University of Michigan. His research interests in evolution focus primarily on speciation and the evolution of ecological interactions among species. He has been a Guggenheim fellow and a Fulbright fellow, president of the Society for the Study of Evolution and the American Society of Naturalists, and editor of *Evolution*. He was elected to the National Academy of Sciences in 2006. He is the author of the successful textbook *Evolutionary Biology* (Sinauer Associates, 1986). Most of his work has centered on the population biology of herbivorous insects and the evolution of their affiliation with host plants. Recent work has focused on the evolution of host specificity, on whether or not constraints on genetic variation are likely to have influenced the phylogenetic history of host associations in a group of leaf beetles, and on the pattern of speciation in this group. Dr. Futuyma's students have worked on diverse evolutionary and ecological studies of insect-plant interactions and of speciation in insects.

James Griesemer is a professor and chair of the Department of Philosophy, University of California, Davis, and a member of the UC Davis Center for Population Biology, the Science and Technology Studies Program, and the Konrad Lorenz Institute for Evolution and Cognition Research in Austria. He received his A.B. in genetics from the University of California, Berkeley, and his M.S. in biology and Ph.D. in conceptual foundations of science from the University of Chicago. His research interests include the history, conceptual foundations, and social organization of genetics, ecology, developmental biology, and evolutionary biology. Dr. Griesemer is president of the International Society for the History, Philosophy, and Social Studies of Biology.

Leroy E. Hood is president of the Institute of Systems Biology. He received his M.D. from the John's Hopkins University School of Medicine and his Ph.D. in chemistry from the California Institute of Technology. His research has focused on the study of molecular immunology, biotechnology, and genomics. His professional career began at Caltech, where he and his colleagues pioneered four instruments—the DNA gene sequencer and synthesizer and the protein synthesizer and sequencer—which comprise the technological foundation for contemporary molecular biology. In particular, the DNA sequencer has revolutionized genomics by allowing the rapid automated sequencing of DNA, which played a crucial role in contributing to the successful mapping of the human genome during the 1990s. In 1992, Dr. Hood moved to the University of Washington as founder and chairman of the first cross-disciplinary biology department, the Department of Molecular Biotechnology. In 2000 he co-founded the Institute for Systems Biology in Seattle, Washington, to pioneer systems approaches to biology and medicine. Most recently, Dr. Hood's lifelong contributions to biotechnology earned him the prestigious 2004 Association for Molecular Pathology Award for Excellence in Molecular Diagnostics. He was also awarded the 2003 Lemelson-MIT Prize for Innovation and Invention, the 2002 Kyoto Prize in Advanced Technology, and the 1987 Lasker Prize for his studies on the mechanism of immune diversity. He has published more than 600 peer-reviewed papers; received 14 patents; and coauthored textbooks in biochemistry, immunology, molecular biology, and genetics; and he is a member of the National Academy of Sciences, the American Philosophical Society, the American Association of Arts and Sciences, the National Academy of Engineering, and the Institute of Medicine. Dr. Hood has also played a role in founding numerous biotechnology companies, including Amgen, Applied Biosystems, Systemix, Darwin, and Rosetta.

David Julius is a professor in the Department of Cellular and Molecular Pharmacology, University of California, San Francisco. He is interested in the molecular biology of sensory transduction and neurotransmitter action in the mammalian nervous system. Dr. Julius is a leading neuropharmacologist whose group has cloned and characterized a number of neurotransmitter receptors and ion channels from the mammalian nervous system. These include temperature-activated channels that also serve as receptors for capsaicin, the pungent ingredient in chili peppers, or for menthol. One of his goals is to understand the molecular basis of somatosensation—the process whereby we experience touch and temperature—with an emphasis on identifying molecules that detect noxious (pain-producing) stimuli. He is also interested in understanding how somatosensation is altered in response to tissue or nerve injury. Dr. Julius received his Ph.D. in biochemistry from the University of California, Berkeley, and his bachelor's at MIT. He is a member of the National Academy of Sciences.

Junhyong Kim is the Edmund J. and Louise W. Kahn Term Endowed professor of biology at the University of Pennsylvania, with joint appointments in the Department of Computer and Information Science and the Penn Center for Bioinformatics. He is also co-director of the Penn Genomics Institute. He received his B.S. from Seoul National University and his Ph.D. from the State University of New York. His current focus is on genomics, computational biology and evolution, and biotechnology. Current projects include neurogenomics, RNA measurement technology, phylogenetics, and yeast comparative genomics. His work includes analyzing the macroevolution and mutational dynamics of the transcriptome in *Drosophila*. He has developed computational tools to study and visualize the timing of transcription in yeast in order to learn about how various transcripts are coordinated to drive the cell cycle. He works on improving the underlying mathematics used for phylogenetic prediction and how to mine the haphazard phylogenetic information from GenBank. He has 20 years of experience in computational biology. He is an associate editor for *IEEE/ACM Transactions in Computational Biology and Bioinformatics*, a member of the editorial board of *Molecular Development Evolution*, and a member of the Scientific Advisory Board of the National Center for Evolutionary Synthesis. He also received the Sloan Foundation Young Investigator Award.

Karla A. Kirkegaard is professor and chair of the Department of Microbiology and Immunology at Stanford University School of Medicine. She received her Ph.D. in biochemistry from Harvard University and has been investigating the genetics, biochemistry, and cell biology of poliovirus and other positive-strand RNA viruses since her postdoctoral work with David Baltimore from 1983-1986 at the Massachusetts Institute of Technology. Dr. Kirkegaard was a faculty member in the "RNA World" of the University of Colorado, Boulder, from 1986 to 1996, where she was an assistant investigator of the Howard Hughes Medical Institute, and a Searle and a Packard scholar. Upon moving to Stanford in 1996, her laboratory's interests have included the biochemistry of RNA-dependent RNA polymerases and immune evasion mechanisms used by positive-strand RNA viruses. In 2006 she was a recipient of a NIH Director's Pioneer Award for her studies of drug resistance in RNA viruses.

Jane Maienschein is regents' professor, president's professor, and parents association professor in the School of Life Sciences and director of the Center for Biology and Society at Arizona State University. She received her Ph.D. from Indiana University. Dr. Maienschein specializes in the history and philosophy of biology and the way that biology, bioethics, and biopolicy play out in society. Focusing on research in embryology, genetics, and cytology, she combines detailed analysis of the epistemological standards, theories, laboratory practices, and experimental approaches with the study of the people, institutions, and changing social, political, and legal contexts in which science thrives. She enjoys teaching and is committed to public education about biology and its human dimensions. Dr. Maienschein has received numerous faculty and teaching awards and has coedited a dozen books and written three books, including most recently *Whose View of Life? Embryos, Cloning, and Stem Cells* (Harvard University Press, 2003).

Eve E. Marder is the Victor and Gwendolyn Beinfield Professor of Neuroscience in the Biology Department and Volen Center for Complex Systems at Brandeis University. She received her Ph.D. in 1974 from the University of California, San Diego. Dr. Marder has studied the dynamics of small neuronal networks using the crustacean stomatogastric nervous system. Her work was instrumental in demonstrating that neuronal circuits are not "hard-wired" but can be reconfigured by neuromodulatory neurons and substances to produce a variety of outputs. Together with Larry Abbott, her laboratory pioneered the "dynamic clamp." Dr. Marder was one of the first experimentalists to forge long-standing collaborations with theorists and has for almost 15 years combined experimental work with insights from modeling and theoretical studies. Her work today focuses on understanding how stability in networks arises despite ongoing channel and receptor turnover and modulation, both in developing and adult animals. Dr. Marder is a fellow of the American Association for the Advancement of Science, a fellow of the American Academy of Arts and Sciences, and a trustee of the Grass Foundation. She was the Forbes Lecturer at the Marine Biological Laboratory in 2000 and the Einer Hille Lecturer at the University of Washington in 2002. She was elected to the National Academy of Sciences in 2007.

Carlos Martinez del Rio is a professor in the Department of Zoology and Physiology at the University of Wyoming. He received his B.Sc. from the Universidad Nacional Autonoma de México and his Ph.D. from the University of Florida. He is a functional ecologist who tries hard to establish connections across disciplines. He studies mutualisms like pollination and seed dispersal and approaches research problems from a variety of perspectives, from the molecular to the biospheric. He and members of his laboratory investigate three broad areas: ecological and evolutionary physiology, stable isotopes as tracers in biological systems, and the spatial ecology of ecological interactions. He was the recipient of the NSF Young Investigator Award in 1992, and he has been an Aldo Leopold fellow since 2004. He has served on review panels and advisory groups at National Science Foundation and several conservation organizations. He is the author of one book and over 100 publications.

Joseph H. Nadeau received his Ph.D. in population biology from Boston University in 1978. He was a postdoctoral fellow with both Jan Klein in the Immunogenetics Department, Max Planck Institute for Biology, Tübingen (1978-1980) and Eva Eicher at the Jackson Laboratory (1980-1981). He was appointed associate staff scientist (1981-1985), staff scientist (1985-1991), and senior staff scientist (1991-1994) at the Jackson Laboratory and then professor in the Department of Human Genetics at McGill University and medical scientist in the Department of Medicine at Montreal General Hospital (1994-1996). He is currently James H. Jewel Professor and chair of the genetics department at Case Western Reserve University School of Medicine and co-director of the Center for Computational Genomics and Systems Biology. He has over 200 research publications. He was a founding member of the International Mammalian Genome Society and a founding editor of *Mammalian Genome*. He was founder and director of the Mouse Genome Informatics Project (1989-1994), founder of the Mouse Gene Expression Database Project (1992-1994), founding editor of *Systems Biology Reviews*, and founder and first director of the Ohio GI Cancer Consortium. He has served on review panels and advisory groups at the National Institutes of Health, the National Science Foundation, and the Human Genome Database. He has organized nearly 40 courses, workshops, and conferences. He has consulted for GlaxoSmithKline, Pharmacia, Celera Genomics, Exelixis, NineSigma, and CellTech Chiroscience and is on the Scientific Advisory Board of Galileo Genomics.

Joan Roughgarden is a professor in the Department of Biological Sciences at Stanford University where she has taught since 1972. She founded and directed the Earth Systems Program at Stanford and was awarded for service to undergraduate education. Dr. Roughgarden has studied co-evolutionary models that combine ecology with population genetics, and her current research focuses on the mathematical theory of reproductive social behavior and applying the cooperative game theory of bargaining and side payments to explain animal social dynamics, especially mating behavior. In addition to a seminal ecology textbook written with Paul R. Ehrlich, Dr. Roughgarden published a 2004 challenge to certain tenets of sexual selection titled *Evolution's Rainbow* (University of California Press,

2005). She received a B.S. in biology and an A.B. in philosophy from the University of Rochester in 1968 and a Ph.D. in biology from Harvard University in 1971. She is the author of five books and over 120 articles.

Julie A. Theriot is an associate professor of biochemistry and microbiology and immunology at the Stanford University School of Medicine. She received her Ph.D. from the University of California, San Francisco. Dr. Theriot studies the transformation of chemical energy to mechanical energy in cell movement. Her work focuses on understanding the mechanisms of actin-based movement of the intracytoplasmic pathogenic bacteria *Listeria monocytogenes* and *Shigella flexneri*. She is investigating these systems at the molecular level to yield insights into the mechanisms of whole-cell actin-based motility and bacterial pathogenesis. Other research interests include establishment and maintenance of bacterial polarity, quantitative videomicroscopy, and image and motion analysis. Honors include a Whitehead fellowship and a Packard fellowship for science and engineering. Dr. Theriot recently received the School of Medicine Award for Graduate Teaching and was named a 2004 MacArthur fellow. She served on the National Research Council Committee on Bridges to Independence: Identifying Opportunities for and Challenges to Fostering the Independence of Young Investigators in the Life Sciences and the Committee on Transforming Biological Information into New Therapies: A Strategy for Developing Antiviral Drugs for Smallpox.

Gunter P. Wagner is the Alison Richard Professor of Ecology and Evolutionary Biology at Yale University and a noted researcher and theorist of developmental genetics and evolution. He was chair of the Department of Ecology and Evolutionary Biology in 1997-2001 and 2005-2008. Prior to joining Yale, he was an associate professor at the University of Vienna. Dr. Wagner and other researchers in his laboratory use mathematical modeling to understand the complex adaptations of organisms, with a focus on the molecular evolution of Hox genes and their role in the origin and early evolution of tetrapod limbs. For example, he has compared the expression of Hox genes between the primitive limbs of salamander and the highly derived limbs of frogs to understand the morphological evolution of these organisms. In another project, his lab studies the evolutionary history of Hox genes in primitive vertebrates and their correlation with the emergence of the developmental body plan of higher vertebrates. He and his team have also developed new mathematical techniques in order to better understand gene interactions and evolutionary biology. Dr. Wagner received his Ph.D. from the University of Vienna. He was awarded a MacArthur fellowship in 1992 and the Alexander Von Humboldt Research Prize in 2005. He is a fellow of the American Association for the Advancement of Science and a corresponding member of the Austrian Academy of Sciences.

Appendix C

Workshop on Defining and Advancing the Conceptual Basis of Biological Science for the 21st Century

Location:
National Science Foundation
4201 Wilson Boulevard
Arlington, Virginia

List of speakers

Robert Full, University of California, Berkeley
Sarah Rice, Northwestern University
Ken Dill, University of California, San Francisco
Garrett Odell, University of Washington
Paul Magwene, Duke University
Melanie Moses, University of New Mexico
Robert Dorit, Smith College
David Hillis, University of Texas, Austin
Virginia Walbot, Stanford University
Nancy Knowlton, Scripps Institution of Oceanography, San Diego
Lee-Alan Dugatkin, University of Louisville
Stephen Lisberger, University of California, San Francisco